SOY PROTEIN AND HUMAN NUTRITION

ACADEMIC PRESS MANUSCRIPT REPRODUCTION

Proceedings of the Keystone Conference on Soy Protein and Human
Nutrition held in Keystone, Colorado, May 22–25, 1978

SOY PROTEIN AND HUMAN NUTRITION

edited by

HAROLD L. WILCKE

DANIEL T. HOPKINS

DOYLE H. WAGGLE
Ralston Purina Company
Checkerboard Square, St. Louis, Missouri

ACADEMIC PRESS   *New York   San Francisco   London*   1979
A Subsidiary of Harcourt Brace Jovanovich, Publishers

ACADEMIC PRESS, INC.
111 Fifth Avenue, New York, New York 10003

United Kingdom Edition published by
ACADEMIC PRESS, INC. (LONDON) LTD.
24/28 Oval Road, London NW1 7DX

Library of Congress Cataloging in Publication Data

Main entry under title:

Soy protein and human nutrition.

   Proceedings of a symposium held May 22-25, 1978, in
Keystone, Colorado.
   1.   Soybean as food—Congresses.   2.   Proteins in
human nutrition—Congresses.   I.   Wilcke, Harold
Ludwig, Date   II.   Hopkins, Daniel T.
III.   Waggle, Doyle H.   IV.   Title: The Keystone conference.   [DNLM: 1.   Soy beans—Congresses.
2.   Vegetable proteins—Congresses.   3.   Dietary
proteins—Congresses.   4.   Nutrition—Congresses.
WB430   K44s   1978]
TX558.S7S65   641.1'2   78-25585

PRINTED IN THE UNITED STATES OF AMERICA

   79 80 81 82   9 8 7 6 5 4 3 2 1

Contents

Contributors

Numbers in parentheses indicate the pages on which authors' contributions begin.

Aaron M. Altschul (369), *Department of Community Medicine and International Health, Georgetown University School of Medicine, Washington, D.C.*

R. L. Anderson (209), *Northern Regional Research Center, USDA, Peoria, Illinois*

C. E. Bodwell (331), *Protein Nutrition Laboratory, Nutrition Institute, SEA, USDA, Beltsville, Maryland*

J. E. Braham (313), *Division of Agricultural and Food Science, Institute of Nutrition of Central America and Panama, Guatemala, C.A.*

Ricardo Bressani (313), *Division of Agricultural and Food Science, Institute of Nutrition of Central America and Panama, Guatemala, C.A.*

Kenneth K. Carroll (261), *Department of Biochemistry, University of Western Ontario, London, Ontario, Canada*

Robert R. Dahlgren (235), *Veterinary Services, Ralston Purina Company, St. Louis, Missouri*

L. G. Elias (313), *Division of Agricultural and Food Science, Institute of Nutrition of Central America and Panama, Guatemala, C.A.*

Samuel J. Fomon (79), *Department of Pediatrics, College of Medicine, University of Iowa, Iowa City, Iowa*

Clifford M. Hardin (5), *Ralston Purina Company, St. Louis, Missouri*

Alfred E. Harper (171), *Department of Nutritional Sciences, University of Wisconsin, Madison, Wisconsin*

Daniel T. Hopkins (235, 299, 355), *Central Research, Ralston Purina Company, St. Louis, Missouri*

M. W. Huff (261), *Department of Biochemistry, University of Western Ontario, London, Ontario, Canada*

G. Richard Jansen (149), *Department of Food Science and Nutrition, Colorado State University, Fort Collins, Colorado*

Constance Kies (325), Department of Food and Nutrition, University of Nebraska, Lincoln, Nebraska

Charles W. Kolar (19), Venture Management Group, Ralston Purina Company, St. Louis, Missouri

Wilda H. Martinez (53), National Program Staff, SEA, USDA, Beltsville, Maryland

John M. McLaughlan (281), Health Protection Branch, Department of National Health and Welfare, Ottawa, Ontario, Canada

D. A. Navarrete (313), Division of Agricultural and Food Science, Institute of Nutrition of Central America and Panama, Guatemala, C.A.

Boyd L. O'Dell (187), Biochemistry Department, University of Missouri, Columbia, Missouri

Joe J. Rackis (209), Northern Regional Research Center, USDA, Peoria, Illinois

Marvin L. Raymond (235), Raltech, Division of Ralston Purina Company, St. Louis, Missouri

D. C. K. Roberts (261), Department of Biochemistry, University of Western Ontario, London, Ontario, Canada

Leonard H. Roberts (359), Protein Division, Ralston Purina Company, St. Louis, Missouri

Nevin S. Scrimshaw (121), Department of Nutrition and Food Science, Massachusetts Institute of Technology, Cambridge, Massachusetts

Fred H. Steinke (307), Central Research, Ralston Purina Company, St. Louis, Missouri

Barbara J. Struthers (235), Central Research, Ralston Purina Company, St. Louis, Missouri

William H. Tallent (209), Northern Regional Research Center, USDA, Peoria, Illinois

Benjamin J. Torun (101), Physiology and Clinical Nutrition, Institute of Nutrition of Central America and Panama, Guatemala, C.A.

John E. Vanderveen (303), Division of Nutrition, Bureau of Foods, Food and Drug Administration, Department of Health, Education and Welfare, Washington, D.C.

Doyle H. Waggle (19), Protein Division, Ralston Purina Company, St. Louis, Missouri

Harold L. Wilcke (1, 377), Ralston Purina Company, St. Louis, Missouri

Vernon R. Young (121), Department of Nutrition and Food Science, Massachusetts Institute of Technology, Cambridge, Massachusetts

Ekhard   Ziegler (79), Department of Pediatrics, College of Medicine, University of Iowa, Iowa City, Iowa

Foreword

The subject of the Keystone Conference, Soy Protein and Human Nutrition, should be viewed from a perspective of the needs and the demands of the world's food consumers. In broad terms, the world's consumers need expanding food supplies at stable prices. They demand foods in forms common or traditional to their specific culture. New foods provided in uncommon forms have been accepted slowly, if at all, in any area of the world. Foods and food consumption are basic components of human culture. Change occurs very slowly. Thus, new sources of food can expand the world supply only when the new source is nutritious and is in a form that will maintain traditional consumption patterns.

There has been a widespread concern for many decades about the food supply for a growing world population. Malnutrition and starvation are basic worldwide problems and concerns. Proper diet and food costs become increasingly interdependent as many people strive to move up the "food ladder." Indeed, we are witnessing the development of a new phenomenon and that is the growing evolvement of food into the very fabric of economic and political affairs of nations.

In the United States, the reporting of the Consumer Price Index and the seemingly unending spiral of increasing costs of the family's food basket are of such concern that it is national and even international news. The very stability of economic and political life is in part dependent upon relatively stable food prices. Achievement of this stability in food prices and particularly those of processed foods can in part be realized through improved efficiency and productivity in all sectors of the food chain.

In recent years, we have seen the growing use of an alternative and economic food protein ingredient in processed foods in the industrialized world—protein derived from the soybean. Technology has been developed and is employed on a commercial scale, which refines or produces soy proteins with a variety of functional properties. Soy protein isolates are utilized in a broad spectrum of foods without changing the form or taste of those foods. The expanding use of these isolated or refined proteins requires the understanding of the nutritional aspects of soy protein by food manufacturers, marketers, and governmental regulatory bodies.

It is vital that the food industry throughout the world develop and utilize new technology for providing nutritional, wholesome, and economic foods. To encourage the development and optimal utilization of new technology, we must complete the research work needed to measure the nutritional contributions of food sources to the human being. Establishment of the methodology for measuring nutritional sources of human beings will allow the governmental regulatory bodies to readily permit the use of new food sources in their food industries. Such developments will provide significant benefits to the consumers in these developed economies. Equally important, however, will be the leadership demonstrated by such progressive new approaches to food regulation and human nutritional needs. The actions of the regulatory agencies of the developed nations are especially critical because they can be the regulation model for other countries of the world.

It is hoped that the research activity of the type that is being reported here, the scientific interaction, and the peer judgments can provide a partial backdrop and springboard for creating the climate to stimulate greater utilization of scientific knowledge and technology in meeting the needs and demands for food throughout the world. Understanding human nutrition and the contribution that soy proteins can make to a nutritious diet for people can be one dimension in the goal of employing new technology to meet the present and future food needs of the world.

P. H. Hatfield

Preface

The soybean, long a very important food in many oriental cultures, has only comparatively recently received serious and widespread interest as a source of food in the United States, even though it has been one of the leading agricultural crops in this country for many years. This is a crop that produces a seed, which not only furnishes an oil that has become our leading food oil but also a protein that is high in lysine and therefore a very good supplement for cereal grains. New technologies have made it possible to produce soybean flours, concentrates, and isolates with very diverse functional properties that fill many needs in food systems. These attributes have persuaded many of our universities, government research institutions, and industrial laboratories of the need for a thorough evaluation of the soybean as a human food. Current results from much of this work are brought together in this volume.

Results reported here from actual nutritional experiments, where soy proteins, particularly the isolates, were fed to infants, growing children, and young adults, have clearly demonstrated that the soybean is a much more nutritionally adequate protein than had been expected on the basis of research done with animals. This has demonstrated a need for more adequate methods of evaluating proteins for the human. A panel of eminently qualified scientists has addressed this question and has made some positive suggestions on better approaches to the evaluation of proteins for humans. Their discussions are presented as a part of this volume. Further, the work reported on the possible effects of the plant proteins on the plasma cholesterol levels of humans opens up an entirely new dimension for the evaluation of proteins. Therefore, this volume will be of particular interest not only to those who are seeking information on the characteristics of the soybean as a food and its potential in the food system but also for those who have the broader interest in how to evaluate proteins for humans.

This volume has been organized to present the present state of knowledge regarding the nutritional values of the soybean as a potential in our food supply, and at the same time to highlight those areas where additional research might be fruitful. It will be of interest to regulators because the positive effects of processing and technology have been clearly delineated in the papers and the discussions. In the words of one of the regulators: "This conference has represented a giant step in the direction of creating that data base; however, it is obvious that much more work remains to be done." This is the primary purpose of this volume.

The editors would like to express appreciation to all of those who prepared papers so thoroughly and painstakingly, and to all those who participated in the discussion. We would also like to express our sincere appreciation to Dave Stone, Conference Coordinator; Bob Green, Manager, Meeting and Travel; Walter Lloyd, Account Coordinator, Traveler's Service; and the staff at Keystone for their very professional assistance in organizing and managing the actual conference; Sue Kowalski for her assistance at the conference; Kris Stein, Pat Leith and Debbie Wamser for their competent assistance in planning and handling the details of arrangements for the conference, registration, and finally in assisting with the manuscripts; the work of the Word Processing Department at Ralston Purina for the preparation of manuscripts, and to Doris Marshall for her meticulous work in preparation of the index.

Harold L. Wilcke
Daniel T. Hopkins
Doyle H. Waggle

Inoue and Taniguchi   Senti and Hardin

Carroll   Bodwell   Martinez   Roberts   Hardin

Jansen   Hatfield   McLaughlan   Hopkins   Altschul

Keystone Lake Village

Saperstein   Informal Discussions
Tallent   Wilcke   Harper   Struthers   Scrimshaw
Steinke   Bressani   Kies   Vanderveen   Torun
Keystone Lodge

Is it going all right?   Presentation
O'Dell   Saterlee
                          Registration
Waggle   Fomon
Snake River and Condominiums

CONFERENCE OBJECTIVES

H. L. Wilcke

This is the first Keystone Conference on soy protein and human nutrition. We hope it will not be the last, but the answer to that question lies largely in the outcome of this first one. I am sure we would all agree that there is an over proliferation of conferences on all subjects in the scientific world. Therefore, unless this Conference can make some specific contributions toward the attainment of the goals we have outlined, or alternately, contribute in some manner that we may not have recognized previously, there would be no particular need for following this conference with another. However, if we attain some degree of success in making "nique contributions to the very important problems under consideration by this group, future conferences should be worthy of consideration.

The objectives of this Conference may be stated as follows:

1) To present the most up-to-date information on the role that soy protein and plant proteins, in general, should fulfill in the human diet. The literature has become increasingly voluminous on various aspects of the nutritional values of plant proteins, the problems associated with the use of plant proteins, particularly when they are substituted for those of animal origin, and some of the limitations for the use of this type of food. Our first objective, then, is to bring together as much as possible all phases of the information regarding the soybean so that we may take a comprehensive look at the extent to which the soybean, which has served so admirably as a source of edible fat and as an excellent source of feed for animals, should be utilized in the human diet.

2) A second objective is to stimulate interest among both the public and private sectors in pursuing research programs directed toward the determination of the proper place of plant proteins in the human diet. The extensive research programs that have been carried out over a period of many years,

together with the information gleaned from the successful use
of soy products in certain oriental cultures, have produced
conclusive data on the excellent nutritional value, the safety
and the desirability of soy products, both protein and oil.
But such products must not be regarded as substitutes in the
food system - they must have unique properties of their own.
Some of these attributes have been elucidated, others are
beginning to surface, and the discussions here, both formal
and informal, will certainly present challenges to all
scientists interested in research on plant proteins, whether
their fields of endeavor may be in the academic, the public,
or the private sectors.
3) The third objective is to stimulate interest in better
methods of evaluating the quality of plant proteins for human
nutrition. Present methods of evaluation, largely based on
animals as they are, do supply a wealth of good information on
both the nutritional value and the safety of food products.
However, the methods that are available to us are admittedly
inadequate. There is always a question when we must extrapolate information from one species to another.

The question of the proper methods of evaluating proteins
for the human diet has been the subject of many discussions.
In this section of the Conference the emphasis will be shifted
from that based almost entirely on the soybean and plant proteins to one of consideration of all proteins.

The question of evaluation is of tremendous importance to
human nutrition and, as such, we feel that the full force of
both the public and private sectors should be brought to bear
in the development of methodology which will be effective in
determining the true nutritional value of all proteins. Only
when such methodology is known and tested will it be accepted
for general use. Thus, this program has been planned to present some of the results that have been obtained with actual
utilization of soy protein products for the direct nutrition
of humans, the successes and problems that have been encountered, the needs and opportunities; and finally, just how
these results may be evaluated properly. The information
presented at this Conference should provide some real incentive to do more than just criticize the methods that are
available to us today. It is hoped that the discussions will
stimulate latent ideas that may be brought forth for collaborative testing, or that the discussions in themselves will
suggest new approaches to this problem, and that plans may be
formulated to develop appropriate methodology to obtain the
information we want and need.
4) The fourth objective is to provide another forum for the
acceleration and improvement of communication of newer and
more complete information on soy protein. The objectivity of

industry scientists is often questioned, and in some quarters
there is a reluctance to accept the conclusions drawn from
industrial research. It should be self-evident that accurate
interpretation of results is nowhere more important than in
business, because sound and lasting business can be built only
upon verified facts. The private sector researcher is dealing
not only with his reputation, but with his very livelihood.
Confidence in this concept can be gained only through
communication and exchange of methodology and results. The
private sector is increasingly publishing research reports in
the scientific journals, although admittedly more of this
should be done.

The question will undoubtedly be rasied as to why this
Company would sponsor such a Conference. How do we justify
the expenditures of shareholders money organizing a public
scientific conference that will provide a comprehensive summary of the information, when the usual policy is not to share
such information, at least not with competitors? In the final
analysis, the ultimate goal of everyone working with the soybean as a food is to provide the information that is necessary
for the full recognition of the commodity as an important food
source. This can only be done when the consumer, the food
industry, and the regulatory agencies are aware of, and
accept, all of the attributes of the product.

This conference is one of the checkpoints in the comprehensive research programs, both public and private, which have
been in progress for a number of years with the objective of
using the soybean as human food. The Ralston Purina Company
has carried on extensive in-house research programs and has
also supported some of the research in the public area. With
this background, it is clear that this Company has a very deep
interest in taking this opportunity to evaluate the present
state of the knowledge on soybean protein.

We hear many recommendations, such as those from the U.S.
Senate Select Committee on Nutrition and Human Needs, to the
effect that we should reduce our consumption of animal products, which are important sources of good quality protein in
our diets. If this recommendation should be accepted, it
follows that we must find alternate sources of high quality
protein. We must not only find other sources of protein, but
we must know what other elements in the diet we are changing
when we lower the consumption of animal products and turn to
the plant sources. The state of the knowledge, in several of
these areas, will be discussed here at this Conference. The
discussions here will not exhaust the possibilities of the
need for information on various nutrients. This program has
been developed, however, to answer as many questions as
possible on just how the soybean qualifies as one of those
alternate protein sources.

IMPACT OF PLANT PROTEINS IN WORLDWIDE FOOD SYSTEMS

Clifford M. Hardin

As an introduction to this discussion of the impact of
plant proteins on various food systems, I want to emphasize
that it is my conviction that there are sufficient food-producing resources and technology in the world today to provide
for the feeding of whatever number of people may live in the
world in the year 2000, better than mankind has ever been
fed. This statement is not a prediction, but rather a statement of potential that can be realized if enough of the right
things are done by enough of the right people.

Food is more than a source of sustenance and nutrition.
Food is intimately intertwined with the cultures and tradi-
tions of peoples everywhere. Certain foods are traditional;
some have symbolic meanings; others are associated with social
status; some are regarded as staples while others are served
when special guests are present; some foods are preferred by
people of one culture and disliked or even tabooed by others;
certain foods are associated with special days, celebrations
and religious rites.

Food is, therefore, more than fat, carbohydrates, protein,
vitamins, minerals and fiber. Food is palatability, taste,
form, appearance and goodness. Food is also a political and
economic reality in almost every country in the world. Food
policies can determine the success or downfall of governments. Food politics will continue to be a hot item of debate
in United Nations and other international forums.

Because food is all of these things, because it is so much
a part of people's daily lives, eating habits and food systems
are difficult to change. But people do change their eating
patterns! Yes, they do, but typically it is more a gradual
evolvement than a sharp shift. Generally, it is an evolvement
to something that is regarded as better. Generally, it is an
evolvement that is associated with rising affluence. But if
improvement in nutrition is to be lasting, if it is to bring
with it improved quality of life as viewed by the partici-

pants, it must be done with appropriate consideration of the
likes, dislikes and cultures of the people involved.

But within all of the diverse patterns, habits and tradi-
tions that must be recognized and taken into account, there
are some valid generalizations that can be made.

As individual incomes begin to rise, there is an almost
automatic and immediate demand for more and better foods from
those who have the money -- whether they live in developed or
developing countries, and whether they live in Europe, Africa
or Asia.

As income levels increase, people start climbing what has
been termed the "food ladder." People with the lowest incomes
live typically on diets that are high in starch - rice, corn,
root crops. Such people crave fat in their diets, and they
buy it when they can afford it. Next, they want protein, including meats.

And, finally, they want some of the more luxury-type items
- fruits and vegetables out of season, and many of the refined
types of foods that are sold in the modern supermarkets in the
Western World.

This succession of food preferences seems to exist with
peoples of all ethnic and geographic backgrounds and all levels of economic development. And therein lies the special
charm of the isolated or refined vegetable proteins. They can
be utilized in a wide variety of food systems to improve nutritive values and functional qualities without displacing the
system itself. They are being increasingly utilized as alternative food proteins in existing food systems. Generally,
they are regarded as additions to the food supply rather than
as displacers of any significant food items. Clearly, they
provide a way for people to move up the "food ladder" without
disturbing familiar and traditional habits of eating.

Take for example, Japan -- whose staple food historically
was rice. As the Japanese economy has grown and Japan has become one of the industrial giants, individual incomes have increased, and the Japanese people are climbing the rungs of the
food ladder in a predictable pattern. First, following World
War II when they were surviving on rice, they greatly increased their consumption of vegetable oils, partly through
massive imports. Then they increased their use of vegetable
protein feed ingredients and began to develop a broiler and
livestock industry. They have become the world's largest im-
porters of soybeans and feed grains. More recently the Japanese have become interested in the use of isolated or refined
soy proteins to extend their supply of fish paste products,
such as kamaboko, at a time when fish supplies are reduced due
to the imposition of the 200-mile fishing limits. They are

also increasing their use of these soy proteins in processed
meat products.

We are seeing this same rising demand for animal protein
in both Western and Eastern Europe and in Russia, and this
lies behind the growing import demand for soybeans and feed
grains. It also explains the growing interest in the Eastern
European countries in the importation of soy isolates for use
in extending their supplies of sausage products. They have
learned that they can increase their sausage consumption with
less expenditure of their limited supplies of foreign exchange
than if they imported the grain and soybeans and increased
livestock numbers. Less developed countries as a group, likewise, are increasingly supplementing their food supplies with
imports.

But no discussion of world protein can take place without
relating it to the total world food supply and total world
food needs.

In recent years, only a handful of countries have produced
more food than their own populations have chosen to consume.
Except for a few specialty crops like sugar, coffee, spices
and palm oil, most food and feed products that move in international trade originate in the United States, Canada, Austra-
lia, Brazil, New Zealand and Argentina. Most of the international movement of food has been to the other developed countries who had access to sufficient foreign exchange that they
could enter the world market and buy what they wanted or ʻ
needed.

Over the period of 20 years that began in 1950, farm exports from the United States rose about 5 percent per year.
In the decade of the 1970's, total world commercial demand has
been growing at a more rapid rate, and it now appears that the
potential exists for farm exports to experience annual increases during the next decade that might average as much as 6
or 7 percent, calculated in constant dollars. While population growth is a factor, the major force in the growing commercial demand for food is rising affluence. Whether people
live in developed or developing countries, or whether they
live in Europe, Africa or Asia, if incomes rise, so does the
demand for food, especially for foods and feeds that will add
to the availability of proteins. Further adding strength to
world markets is the apparent decision of Russia and some
other countries to depart from their traditional pattern in
short crop years -- that of tightening their belts and tough-
ing it out. Their pattern now seems to be to enter world markets and buy, rather than cut back heavily in consumption.

The import potential of the Peoples Republic of China remains generally unknown. Their leaders have indicated that
they expect China to be a "full participant" in the industrial

world by the year 2000. With that kind of objective vigorous-
ly pushed, China may also become a major importer of food.

We need to add to this demand the continuing purchases
that will be made by the PL 480, Food for Peace Program, the
purchases for food aid by other countries and the purchases
for relief feeding by the various United Nations groups.

It is my judgment that the American farmer will be able
during the 1980's to produce enough to satisfy at reasonable
prices the rising worldwide commercial demand for the crops we
grow for export -- at least in most years. We must recognize,
however, that because of the vagaries of weather, there will
continue to be shortages of some crops in some years and surpluses in others.

It is possible that by the end of the 1980's, we will be
straining our production capabilities. Much will depend on
our ability to continue to increase yields, on whether price
and profit opportunities will cause additional but less productive land to be utilized, on costs of energy and other production inputs and on the general availability of water for
irrigation purposes.

But let us turn to the "Other World."

Two thirds of the world's people live in developing countries with burgeoning populations. Malnutrition is still
widespread and the gap between the "haves" and the "have-nots"
is still large. In fact, the gap may be widening in a number
of countries.

The United States and other developed countries simply
cannot begin to produce enough food to meet both the commercial demand and the real nutritional needs in the world. They
could not produce that much food even assuming some magic way
could be found to finance it. If starvation and malnutrition
are to be stemmed or prevented, the developing countries simply have to learn how to produce more on their own soil and
provide the means of getting it to the people who need it.
There is no other way.

But, wouldn't it help, really, if we in the United States
were to reduce our consumption of meat and release grain for
consumption in the developing world? The answer is, no! To
the extent that we reduced the commercial demand for grain and
lowered prices, we would be signaling to farmers to reduce
output in future years.

I recall vividly in late 1971, when we still had large
surpluses of grains as we do today, of discussing whether any
way could be found to get those surplus stocks to people who
needed them -- and, beyond the PL 480, Food for Peace Program,
and some of the special church programs, there was no way.
There still is no way unless food aid can be expanded, even
though today, we have large surpluses and prices are low.

Hopefully, either through some of the United Nation's spon-
sored programs or directly, other developed countries and some
of the O.P.E.C. countries will increase their financial participation in relief feeding programs to the end that, collec-
tively, we can be more effective in responding to famines and
other catastrophes on an emergency basis. Hopefully, also, we
will be able to convince some of our importing customers to
build storage facilities on their own shores, fill their bins
in years like this one, and even out their own demands to the
end that pressure on the market in short crop years will be
less severe.

It is technically correct to say that more people can be
fed from crops grown on an acre of land when the crops are
consumed directly than when the crops are fed to livestock.
Even so, I intend to demonstrate that the existence of a
strong livestock industry in 1974 actually helped to alleviate
the world grain shortage in that period, and that in the future, the United States livestock feeding industry should itself be regarded as an effective part of a world grain reserve, and an aid to the maximizing of grain exports -- and
especially so in years of world shortage.

But let us go back to 1974 and examine what happened.

Figure 1 shows exports of feed grains and wheat between
1970 and the crop-marketing year that just ended. You will
note that the increase in feed grain exports has been dramatic
-- going from about 21 million tons in 1970-71 to 56 million
tons last year. The growth in exports of wheat are not dramatic, but they are still up on a trend basis by about 5 percent a year. You will note also the modest drop in exports of
both feed grains and wheat in the year following the short
crop in 1974.

Total feed grain production in the United States in 1974
was down *17 percent* as indicated in Table 1. Now let us
examine how the United States livestock industry responded to
this shortfall. Between December, 1974 and November, 1975,
the pig crop was reduced by *15 percent* from the previous
year. By January 1, following the short harvest, the number
of beef cattle on feed was reduced by *26 percent* from a year
earlier and by April 1, 1975, further reduced to effect a *31
percent* decrease from the preceding year. Total feed grain
use in this country from harvest to harvest was actually reduced by *24 percent*. Yet exports of feed grains were down
by only *10 percent*. Clearly our feeders did adjust quickly
and effectively, and because they did, the impact on the rest
of the world was less severe than it otherwise would have
been. Incidentally, wheat exports were cut more severely than
feed grains, perhaps partly because there was little wheat
being fed to livestock and there was, therefore, no livestock

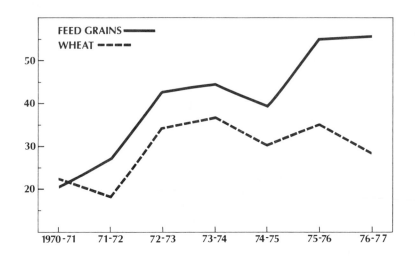

FIGURE 1.   U.S. Exports of Feed Grains and Wheat
(Marketing year - 1 million tons).

buffer to draw on.
     "But", someone may say, "if you hadn't had all that live-
stock in the first place, we could have fed still less and
helped the world more." Again the answer has to be, no! We
expanded our grain producing base in this country in response
to a growing consumer demand for meat and other animal products. In the absence of that kind of strong and continuous
demand for grain to feed livestock, the acres devoted to feed
grain production would have been much smaller. We would have
had the same weather, fewer livestock to take grain away from,
and our contribution to the world grain shortage would have
been significantly less.
     Moreover, without our large livestock population, espe-
cially the ruminant animals, we would not be able to convert
the tremendous quantities of pasture, forage and other coarse
materials that are available in this country into human food.
Also, ruminants can be shifted quickly to roughage feeds in
times of grain shortages or high grain prices. In other
words, *they act as a "surge tank" in the food line.*
     It works the same way in a developing country. Since Bib-
lical times, animals have been used as a buffer against crop
failure. Professor Donald Paarlberg, writing in 1969 on this
subject, states as follows: "A big adjuster is livestock ---

Table 1.   Grain and Livestock - Production and Exports
*   1974-75 as Percent Change From Year Earlier*

Feed Grain Production, 1974   - 17%

Pig Crop (Dec.-Nov., 1974-75)   - 15

Cattle on Feed
       January 1, 1975   - 26
       April 1, 1975   - 31

Feed Grains Fed In U.S.[1]   - 24

Exports of Feed Grains[1]   - 10

Exports of Wheat[1]   - 16

---

[1]*Marketing year, tons*

If the food supply is reduced, we eat the livestock and then
eat the crops the livestock otherwise would have eaten.   The
potential of this adjuster is enormous.   Not all countries
have this shock-absorber in their food supply.   The United
States has it --- some countries --- have long been so near
the margin of want that the livestock population is very small
and there is little cushion to avert disaster."
      There is evidence that more and more of the developing
countries are adding some livestock to their economies.   Over
the period of the 1970's, feed grain use in the United States
and the other developed exporter nations has actually dropped
(Table 2).   In the same period, there have been significant
increases in feed grain usage in Japan, Western Europe, and
Central Planned Countries.   The largest relative increases
have occurred in the developing countries, especially in Mexico and Central America, South America, North Africa and the
Middle East.
      But doesn't the expansion of a local livestock industry
tend to "sop up" local grain supplies and reduce the quantities of grains that would otherwise be available to the masses
of people with lowest incomes?   Doesn't the existence of a
livestock industry cause a diversion of acres from vegetables
and pulses into grains to provide feed for the livestock?
Yes, these things can happen, but only if the pricing policies

Table 2. Feed Use of Grain

Country/Region	1960/61-62/63	1969/70-71/72	1976/77
	(Million Metric Tons)		
I. Developed Countries	187.9	252.1	246.8
United States	110.8	136.5	117.0
II. Central Planned Countries	77.5	143.8	177.0
III. Developing Countries	17.3	29.5	41.8
IV. Rest of World	- -	- -	- -
V. World Total	282.7	425.4	465.6

Source: World Agricultural Situation, ERS, USDA, July, 1977.

of local governments permit them to happen, and if existing food-producing resources are being utilized completely, which is seldom the case.

Actually, increasing the local supply of animal products seems always to be helpful in improving total nutritional health, and as I have pointed out, in helping people to cope with temporary periods of bad weather and crop shortages. While meat and animal products are purchased initially by those with money, the existence of livestock in the economy seems to be a necessary step in the process of moving to a higher level of nutrition for all the people. This is a fact of economic development that often is ignored by those who approach nutrition with a strictly mathematical type of analysis.

As a further back drop for tonight's discussion of the future role of plant proteins, it seems appropriate to look at the kinds of protein the peoples of the world are now eating, any shifts that may be occurring in production and consumption patterns and the total available supply of proteins in relation to the growth in population.

Table 3 indicates that nearly three-fourths of the protein produced in the world is in the form of grains, oilseeds and

Table 3.   Estimates of World Protein Production, 1976

Protein Source	Million Metric Tons	% of Total
Cereals, roots and tubers	142	57
Oilseeds	31	12
Roots and tubers	9	4
Fruits, vegetables, pulses, nuts	14	6
TOTAL VEGETABLES	196	79
Fish products	14	6
Meats, poultry, eggs	23	9
Milk, cheese, butter	16	6
TOTAL ANIMAL	53	21
GRAND TOTAL	249	100

Source: Economic Research Department, Ralston Purina
Company.

root crops; about 6 percent comes from fruits, vegetables,
pulses and nuts, about 6 percent from fish and about 15 percent is in animal products.

When the protein fed to livestock is removed (Table 4), we
find that about 48 percent of the protein available for human
consumption is in the form of grains and root crops. About 15
percent currently comes from oilseeds, fruits and vegetables,
nuts and pulses; about 7 percent from fish and about 29 percent in the form of meat, poultry and other animal products.

If all 4 billion of us who inhabit the earth today were
given allocations of protein and other food components, like
rats in cages, according to our real physical needs, there
would be sufficient food to supply sound nutrition for every-

Table 4. FAO Estimates of World Average Per Capita Food
Protein Supply, 1961-65 and 1974

Protein Source	Grams/Day*		Percent of Total	
	1961-65	*1974*	*1961-65*	*1974*
Cereals, roots and tubers	32.7	33.4	50	48
Fruits, vegetables, nuts, oilseeds and pulses	10.6	10.2	16	15
Fish	3.6	4.5	5	7
Meat, milk, poultry, eggs and other animal	18.3	19.9	28	29
Other	0.9	0.9	1	1
TOTAL	65.9	69.0	100	100

*Totals may not add due to rounding.

Source: *Monthly Bulletin of Agricultural Economics and Statistics*. Vol. 26, No. 5. Food and Agriculture Organization of the United Nations, Rome, May, 1977.

one and with a comfortable margin of safety. But the world does not work that way, and it will not in the future. Aggregate numbers and averages are helpful in establishing outer limits and in measuring trends; they give us little insight into the special problems of specific groups and especially those who are consuming at less than average and less than adequate levels.

For example, of the total protein consumed by Americans about 70 percent comes from animal sources. While worldwide, it is estimated that 70 percent of the protein consumed comes from plants, and in many societies it exceeds 90 percent.

These data suggest that if the protein content of the major cereal crops, rice, wheat and maize could be increased significantly, a major contribution to the elimination of malnutrition could be accomplished. Polished rice, for example, contains about 6 percent protein. If new varieties containing even 7 percent could be widely introduced, this by itself would add many tons of protein where it is most needed. Varieties of rice which contain as much as 11 percent protein have been developed, but they are not grown generally, because there is a straight line, inverse relationship between yield and protein content. Since rice is sold by weight with no allowance for extra protein, growers stay with those varieties that produce the highest yields.

Although tremendous progress has been made in the development of high lysine corn, it is still generally lower yielding and still has some storage problems that have not been solved.

In the absence of a special bonus for high protein grains or until high protein varieties are developed that yield as well as standard varieties, little if any of the new types will be produced. If the difficulties could be overcome and the transition could be accomplished, the world would take a giant step toward eliminating protein malnutrition.

Any shifts that occur in these overall production and consumption patterns in the next two decades will be modest. They will be principally in response to demand as reflected by prices set in the free market or by prices artificially set and supported through subsidies by central governments. There is at least a basis for hoping that total calories on a per capita basis will rise. Within that pattern, I would predict that during the next two decades, there will be some shift toward greater per capita consumption of both fats and proteins on the part of those with improved incomes.

In my judgment, one of the most urgent needs in the world today is for improved varieties of high protein legumes that are adapted to the tropics.

Dr. Lewis M. Roberts (1970), of the Rockefeller Foundation, states, "A look at world acreage, production, and yield figures for all major food crops reveals that the legumes are far behind in yield per hectare, although the total acreage devoted to them is relatively high, reflecting their importance. With the exception of such crops as beans *(Phaseolus vulgaris)*, peas *(Pisum* spp.), peanuts *(Arachis hypogaea)*, and soybeans *(Glycine max)*, which are of importance in the developed world, the majority of the food legumes are of importance only to the people of developing countries. They have had little economic value as cash crops or as exports, because utilization has been almost solely confined to immediate home consumption, or at best to close-range, farm-to-consumer marketing. As a result, these crops have not attracted attention in improvement programs, in contrast with wheat, rice, or maize, which have always had worldwide importance."

What Dr. Roberts is saying is that the food legumes of the type that are traditional and preferred in the developing countries in the tropics are losing out economically to high yielding grain crops at a time when they are urgently needed in the diets of the local people. There are at least 20 high protein leguminous crops, such as pigeon peas, cowpeas, wing beans, mung beans and chick peas which are preferred and eaten traditionally by various groups of people in the tropics.

The solution, of course, is the generation of an aggressive program of plant breeding designed to increase yields of the legumes so that they can be grown competitively and sold for consumption in the rapidly expanding urban centers. Such a program, accompanied by enlightened government pricing programs, holds the potential for significant improvement in protein availability.

Recently, some sound genetic work on high protein crops for the tropics has been started. For example, both the Rockefeller and Ford Foundations are sponsoring research on selected crops, and work is also underway at several of the international tropical research centers.

Soybeans are grown in a few areas in the tropics, but the yields are extremely low because the soybean is sensitive to day length and has not been adapted for areas near the equator.

There are many things that can be done to improve the food supply for that part of the world that needs it most. I have mentioned increasing the protein levels in cereals, improving the yields of the food legumes and developing a balanced livestock sector. I should also mention the need to improve yields of existing crops, through greater and more efficient use of fertilizer; more effective control of pests; improved storage, processing and distribution systems, and the greater and more efficient use of irrigation.

Perhaps the most important single component of success will be national leadership. What is needed is a degree of enlightenment and political courage that has not always been in evidence. The Green Revolution, contrary to some reports, has been highly successful in every country where local leaders have given it a chance. But too often, country leaders "short term" it by giving in to urban pressures for cheap food. If this happens, and farm incomes drop so farmers can no longer afford to buy fertilizer, seed and water, food production may actually decline.

As programs such as I have listed or discussed begin to take hold in country after country, the isolated or refined vegetable proteins, in their special role as international food proteins, will be used widely throughout the world as critical ingredients in existing food systems. This will happen because these proteins are of high nutritional quality, because they are low cost in relation to other protein sources, because of their functional characteristics, because of their great versatility and adaptability and because their use supplements and does not displace or disrupt existing systems.

It is vital that the food industry throughout the world continue to develop and utilize new technology for providing nutritional and wholesome economic foods. It is also important that we improve the methodology for measuring the nutritional responses of people to the new foods.

Research activity of the type that is being reported at this meeting becomes critical. People who are considering the use of new foods need assurances from competent sources concerning nutritional quality, safety and efficacy.

Scientific interaction and peer judgments can and do provide the backdrop for the design of proper regulatory structures and standards. The actions of U.S. and other Western World regulatory agencies are especially significant, because they frequently become the model for regulations in other countries.

Finally, independent research evaluations provide an important basis for product improvement and further development. There is indeed hope and the potential does exist for mankind to be fed better in the future than in all history.

REFERENCES

Paarlberg, Donald (1969). In "Overcoming World Hunger", (Clifford M. Hardin, ed.). Prentice Hall, Englewood Cliffs, NJ, pp. 42-43.

Robert, Lewis M. (1970). "The Food Legumes - Recommendations for Expansion and Acceleration of Research." The Rockefeller Foundation, NY.

TYPES OF SOY PROTEIN PRODUCTS

D. H. Waggle and C. W. Kolar

The soybean originated in Eastern Asia where it has been utilized as a food source for centuries. It was cultivated for food in China long before written records were kept and was even recommended for its therapeutic value during its early history. One can only speculate on how soybeans were first prepared for human consumption. Miso, a soy paste which is often used as a soup base, is a traditional fermented oriental product which is still popular today. Shoyu or soy sauce is a fermented product which has also been widely accepted in many western countries.

Soybeans are a relatively recent agricultural crop in the United States. Early U.S. interest in the soybean was for the oil which was pressed from beans. Hydraulic and screw presses were first used to separate the lipid fraction from the meal or cake. Much of the oil was used in paint and other industrial applications, and the meal was considered a by-product which was used primarily as cattle feed or fertilizer.

Commercial solvent extraction of the oil was initiated in the early 1930's. Coincidentally, food uses for the oil also started to develop at about that time. Processing of the soybean into food products such as soya flour, soy protein concentrate, and isolated soy protein developed in the 1950's. These forms of soy food products are marketed in a wide variety of forms, including dry powders and texturized products. Efficient extraction of the oil led to the development of refined soy proteins which were used as adhesives in the plywood industry. Today, nearly all the oil goes into food applications while the meal is regarded as a very important protein supplement for livestock and poultry.

SOYBEAN SUPPLY

Western Hemisphere soybean production has grown by a factor of five in the last twenty-five years. The major Western
producing countries today are the United States and Brazil, as
shown in Figure 1. Production in the United States has been
increasing at a rate of approximately 6.5% annually, resulting
primarily from increased acreage, but also, to some extent,
from improved crop yield. Brazil, on the other hand, has been
increasing production at a compound rate of approximately 35%
over the past five years. Brazil's soybean production now approximates one-fourth the size of the United States' crop.
This dramatic rise in soybean production is directly related
to increased prosperity around the world with the resulting
demand for meat and livestock products, which is still the ma-

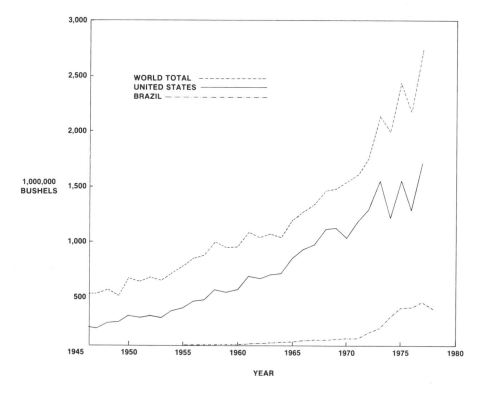

FIGURE 1. World production of soybeans from 1945 to 1977.

jor use of soybean meal. Adoption of modern, efficient feeding practices has boosted world requirements of high protein
feed for expanding livestock production.

Markets for soybean meal are principally high protein supplements for poultry, hogs, and cattle, as well as for other
livestock rations. These markets have been expanding much
faster than those for oil, which is one of the reasons why
soybeans, which have a relatively high meal to oil ratio, are
preferred to other oil seeds. The demand for soybean meal as
a raw ingredient in food processing applications has been
growing rapidly, but this segment still comprises less than 5%
of the total soybean production in the United States.

Agronomists have developed a number of commercially acceptable varieties and no single variety dominates the market.
Soybean varieties are divided into ten maturity groups adapted
for northern latitudes from Southern Canada, to southern latitudes for the southernmost region of the Gulf Coast.

Plant breeding for yield, disease resistance, and composition is a continuous program; therefore, major varieties are
being continuously improved. Segregation of specific varieties is not practiced commercially.

Seed Structure and Composition

Soybeans are typical of legume seeds which differ in size,
shape, and color depending on the variety. They range from
small round beans to large oblong, flattened seeds of yellow,
brown, green, black, or combinations of these colors. Common
field varieties of soybeans which are agriculturally important
are spherical and yellow. Typical soybeans used for food
products are shown in Figure 2. A cross-section of a soybean
(Figure 3) shows the major structural parts; the hull and the
cotyledon. Two minor structures, the hypocotyl and plumule,
are not shown. The hull is made up of an outer layer of palisade cells, a layer of hourglass cells, smaller compressed parenchyma cells, aleurone cells, and finally compressed layers
with endosperm cells. The cotyledon is covered with an epidermis and is composed of numerous elongated palisade-like
cells which contain protein and oil. The bulk of the proteins
are stored in protein bodies which may vary from 2 to 20 microns in diameter within the cotyledon cells. The oil is located in smaller structures, 0.2 to 0.5 microns, called
spherosomes which are interspersed between the protein bodies
(Tombs, 1967 and Wolf, 1975).

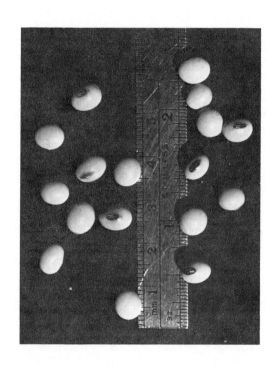

FIGURE 2. Photograph of typical U.S. soybeans utilized in production of food products.

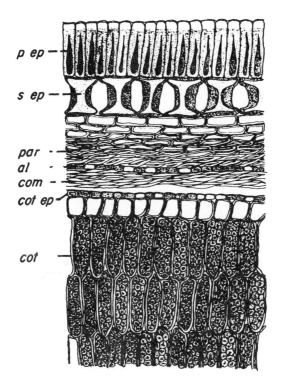

FIGURE 3. Cross section of soybean seed coat and portion
of cotyledon showing: spermoderm, which consists of *p ep*
palisade cells of epidermis; *s ep* hourglass cells, and
par, parenchyma, both of subepidermis; *al*, aleurone cells;
and *com*, compressed cells of the endosperm, *cot ep*, cotyledon epidermis, and *cot*, aleurone cells of cotyledon.

Composition of Source Material

The soybean is composed of three major components, the
hull, cotyledon, and hypocotyl in portions approximately 8%,
90%, and 2%, respectively. Typical values of composition of
the total soybean are presented in Table 1 (Kawamura, 1967).
Constituents of major interest for food applications are protein and oil which are 42% and 20%, respectively. One third
of the soybean is carbohydrates, which include various polysaccharides, stachyose, raffinose and sucrose (Kawamura,
1967). The balance of the materials present in soybeans is
described as ash, which includes many minerals.

Table 1. Typical composition of soybeans[1]

	%
Protein (N x 6.25)	42
Oil	20
Total Carbohydrates[2]	35
Ash	5.0
Crude Fiber	5.5

[1]*Moisture-free basis.*
[2]*Includes crude fiber. (From Kawamura, 1967.)*

Storage and Handling

Soy protein products for human consumption are produced
from high quality soybeans as the raw material. Soybeans are
classified as cereal grains; hence, trading is regulated by
the U.S. Grain Standards Act. Classification of soybeans is
according to color, and yellow soybeans constitute the major
commercial class. Grades are based on test weight, moisture
content, percentage of split, damaged kernels, and foreign material. Table 2 lists requirements for numerical and sample
grades of soybeans (Official Grain Standards, 1970). Soybeans
of grades Number 1 and Number 2 are used for processing food
grade protein products.

Proper receiving, handling, and cleaning of soybeans is
important in producing high quality food products. Cleaning
is accomplished by magnetic separators, screening, and other
techniques such as air classifiication and mechanical separators. It is important to remove as much foreign matter as
possible, in order to maintain a pure product and for processing efficiency. Clean soybeans will retain their quality in-
definitely in bulk storage, providing the moisture content
does not exceed 12%.

The general outline of the process for producing crude oil
and defatted soybean products is shown in Figure 4. The
cleaned beans are cracked into multiple pieces, usually six to
eight, and the hull is loosened by use of corrugated rolls re-
volving at slightly different speeds. For food use dehulling

Table 2. United States Standards for Soybeans

| | | | | MAXIMUM LIMITS OF: | | | |
Grade	Minimum Test Weight Per Bushel Pounds	Moisture (%)	Splits (%)	Total (%)	Heat Damaged (%)	Foreign Material (%)	Brown, Black and/or Bi-Colored Soybeans in Yellow or Green Soybeans (%)
1	56	13	10	2.0	0.2	1.0	1.0
2	54	14	20	3.0	0.5	2.0	2.0
3[1]	52	16	30	5.0	1.0	3.0	5.0
4[2]	49	18	40	8.0	3.0	5.0	10.0

U.S. Sample Grade: U.S. sample grade shall be soybeans which do not meet the requirements for any of the grades from U.S. No. 1 to U.S. No. 4, inclusive; or which are musty, sour, or heating; or which have any commercially objectionable foreign odor; or which contain stones or which are otherwise of distinctly low quality.

[1]Soybeans which are purple, mottled or stained shall be graded not higher than U.S. No. 3.

[2]Soybeans which are materially weathered shall be graded not higher than No. 4.

(Official Grain Standards, 1970)

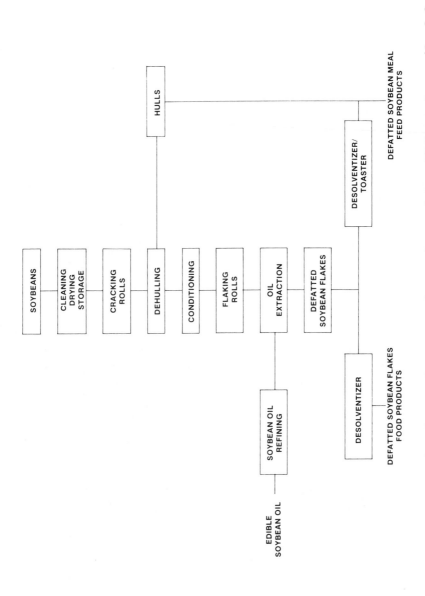

FIGURE 4. Schematic diagram for processing soybeans into edible soybean oil, defatted soybean flakes (food products) and defatted soybean meal (feed products).

is usually practiced after cracking in order to minimize the crude fiber content. Oil extraction efficiency may be improved by removing the hulls, since hulls contain very little oil. Hulls may be removed by aspiration of the cracked beans and collected to yield a product known as mill run and mill feed. Soybean hulls may also be returned to the soybean meal after oil extraction to make a 44% protein meal product.

The cracked beans are conditioned to 10-11% moisture at a temperature of 63-74°C and then flaked by passing them through smooth rolls. The clearances between the rolls are adjusted so that the flakes are a uniform 0.254 millimeters to 0.381 millimeters thick. Flaking ruptures the cotyledon cells in the soybeans and reduces the distance that the oil must diffuse, thereby facilitating extraction with organic solvents in later processing.

Soybean Oil

Soybean oil has been a principal product from soybean processing for many years. In the later 19th century, oil was recovered from soybeans by a pressing operation. While inefficient by today's standards, pressing soybeans yields about 75% recovery of the oil. In the early 20th century, a continuous oil extraction process using an organic solvent to remove oil was developed. The most common solvent used is n-hexane. These continuous extraction systems typically remove about 95% of the oil present in the soybean. Solvent extractors were introduced in the 1930's and some of these early models are still operating.

Currently, extractors employ a variety of ways to contact flakes, including a presoaking period to remove some of the oil in a stationary basket with the pumping of solvent miscella in a progressive stepwise countercurrent flow of flakes and solvent. Miscella is filtered to remove fines and the solvent is stripped from the crude oil by passage through preheaters using thermo-evaporators and stripping columns. The latter are usually packed, and they are steamed countercurrently under diminished pressure to remove the last of the hexane.

The material balance for a typical soybean oil extraction process is shown in Table 3. Soybean oil is primarily used as a starting material for a wide variety of food products such as shortening, margarine, cooking oil, and in salad dressing. A partial list of uses of soybean oil is shown in Table 4.

The most common method of removing solvent from the soybean flakes is the desolventizer-toaster. This equipment recovers the hexane and simultaneously toasts the flakes to obtain optimum nutritive value for production of soybean meal. As the flakes pass through the desolventizer, the temperature

Table 3. Material Balance for Soybean Processing
            60 Pounds = 1 Bushel

	Pounds
Soybean Meal (49% Protein Meal)	43.3
Oil	11
Hulls	4.2
Shrink	1.5

Table 4. Food Uses of Soybean Oil

                        Uses

Frying Fats

Margarine

Mayonnaise

Pharmaceuticals

Salad Dressing

Salad Oil

Sandwich Spreads

Shortening

is gradually increased to 110°C which lowers the moisture content and volatilizes the hexane for efficient removal. After drying and cooling, the flakes are ground into a meal for use in feed applications.

Flash desolventization is a newer process and has the advantage of producing soybean flakes which have high protein solubility properties. Materials from this process generally have a protein dispersibility index (PDI) in the range of 70-90, depending on the processing conditions. Flash desolventizers can also be operated so as to produce products for which the protein dispersibility index value of the soybean flakes will be substantially reduced.

Food Products from Defatted Soybeans

Three general categories of food products are produced from defatted soybeans based on the protein content; soy flours and grits (minimum of 50% protein), soy protein concentrates (minimum of 70% protein), and isolated soy proteins (minimum of 90% protein). Within each general category there are a number of product types with different physical, chemical, and functional properties. The typical composition of soybeans, defatted soy flour, soy protein concentrate and isolated soy protein is shown in Table 5.

Soy Flours and Grits

Soy flours and grits are typically prepared directly from defatted soybean flakes with a minimum of additional processing required. All edible soy grits and soy flours are made from dehulled soybeans. Typical composition of these products is given in Table 6. They contain a minimum of 50% protein,

Table 5. Typical Composition of Soybeans and Soybean Products[1]

	Protein (N x 6.25) (%)	Oil (%)	Total Carbohydrates[2] (%)	Ash (%)	Crude Fiber (%)
Whole Soybean	42	20	35	5.0	5.5
Defatted Soy Flour	54	1.0	38	6.0	3.5
Soy Protein Concentrate	70	1.0	24	5.0	3.5
Isolated Soy Protein	92	0.5	2.5	4.5	0.5

[1]*Moisture-free basis.*
[2]*Includes crude fiber.*

less than 1% fat and less than 3-1/2% crude fiber. The carbohydrate fraction contains polysaccharides, stachyose, raffinose and sucrose as shown in Table 7. Stachyose and raffinose
have been implicated as causative factors of flatus related to
soybeans (Rackis, 1970). The processes used for production of
the various types of flours and grits are outlined in Figure
5.

Table 6. *Typical Composition of Defatted, Dehulled Soy Flour/*
Soy Grits[1]

	%
Protein	54
Oil	1
Ash	6
Soluble Carbohydrates	17
Insoluble Carbohydrates	21

[1]*Moisture-free basis.*

Table 7. *Soluble Carbohydrate Composition of Dehulled,*
Defatted Soybean Meal[1]

Carbohydrate	%
Hexose	Trace
Sucrose	5.7
Raffinose	4.1
Stachyose	4.6
Verbascose[2]	Trace

[1]*Kellor, 1974*
[2]*Kawamura, 1967*

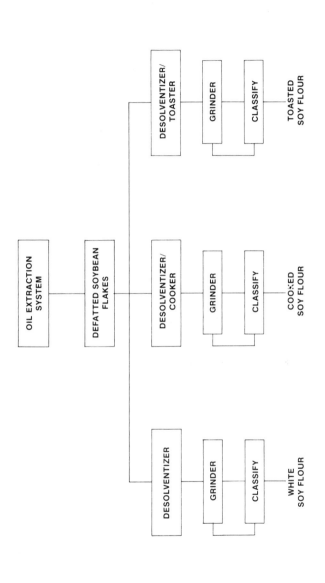

FIGURE 5. Schematic illustrating the processing of white, cooked, and toasted soy flour from defatted soybean flakes.

The range of protein solubility can be 85% and above for uncooked soy flour to less than 15% for toasted soy flour. The uncooked and lightly heat-treated products contain relatively high levels of trypsin inhibitor activity and need additional heat processing to reduce the inhibitor activity in the ultimate food application.

Soy flour and grits are classified by particle size to differentiate the various products.

Product	Mesh Size (U.S. Standard Sieves)
Grits	
Coarse	10–20
Medium	20–30
Fine	40–80
Flours	100 or finer

*U. S. Standard Screens.

Most soy flours are ground to 200 mesh with specialty flours having smaller particle sizes. Grits are obtained by coarse grinding and screening to obtain the specified particle size.

Flour and grits are used to produce a wide variety of products with varying particle size, protein solubility and fat content. The typical composition of commercially available soy flours (dry basis) are given in Table 8. The protein content of these soy flours ranges from 43% to 53% protein, depending on the oil content. The appropriate amount and type of oil or lecithin is added to defatted soy flour to produce refatted (high and low levels) and lecithinated (high and low levels) soy flour products.

Full fat soy flours which are enzymatically active are made by grinding cleaned, dehulled soybean flakes containing 19–21% natural oil. Toasted full fat soy flours can be produced by an extrusion cooking of cracked dehulled soybeans. The soybeans are heated dry to inactivate lipoxygenase, then tempered, and finally extruded. The extruded product is cooled and ground, resulting in a full-fat flour (Mustakas, 1970).

Soy flour can be textured by a number of processes. The two most common methods are the thermoplastic extrusion process (Flier, 1976 and Atkinson, 1978) and steam texturization process (Strommer, 1973). Products from these processes are produced in a wide variety of sizes, shapes, colors, and flavors, depending on the intended food application. Dry expan-

*Table 8. Typical Composition of Commercially Available Soy
          Flour Products[1]*

	Oil (%)	Protein (%)	Carbohydrate[2] (%)	Ash (%)
Defatted Soy Flour[3]	1	54	38	6
Full Fat Soy Flour[4]	20	40	35	5
Refatted Soy Flour[3] (High Fat)	15	45	33.5	6.5
Refatted Soy Flour[3] (Low Fat)	5	48	41.5	5.5
Lecithinated Soy Flour[5] (High Fat)	16.4	48.0	28.6	6.4
Lecithinated Soy Flour (Low Fat)	6	48	41	5

[1]*Moisture-free basis.*
[2]*Calculated by difference.*
[3]*Atkinson, 1978.*
[4]*Kellor, 1974.*
[5]*Smith & Circle, 1972.*

ded products are typically crunchy; however, upon hydration
they become fibrous and chewy in nature.

Soy Protein Concentrates

Soy protein concentrates are defined as the major proteinaceous fraction of soybeans prepared from high quality, sound,
cleaned, dehulled soybeans by removing most of the oil and water soluble nonprotein constituents. Soy protein concentrates
shall contain not less than 70% protein (N x 6.25 on a
moisture-free basis). Following the guidelines of this definition restricts the use of the term concentrate to only those
products which contain more than 70% protein.

Soy protein concentrates are manufactured by removing the
soluble carbohydrate fraction from thoroughly cleaned, defatted soy flakes or soy flour. The process is based on the
principle that the cellulosic flake skeleton does not dissolve

and the protein can be temporarily kept from solubilizing
while most of the sugars, salts and other low molecular weight
components are removed. Mechanisms used to inhibit protein
solubility include: 1) leaching with aqueous and/or organic
solvents with a concentration range in which the proteins are
insoluble, but which extract the nonprotein solubles; 2)
leaching with aqueous acids in the isoelectric range of minimum protein solubility, about pH 4.5.; 3) leaching of cooked
or toasted defatted soy flakes with hot water; and 4) leaching
in the presence of multivalent cations. The first three techniques tend to permanently insolubilize the protein, which inhibits protein functionality. The method of choice to maintain functionality is the acid leach process.

Since all insoluble carbohydrates, including cellular material, remain in the product, it is imperative that little
extraneous matter not originating from the cotyledon of the
bean, be allowed to remain in the flakes. This includes
hulls, hila and the foreign materials, which are detectable
through increased fiber content. Extensive cleaning along
with low residual fat enhances the protein level of the
flakes, which must be inherently high in order to achieve
products with 70% protein.

Three processes are generally used to commercially produce
soy protein concentrates: the aqueous alcohol leach, dilute
acid leach, and the moist heat and water leach. The processes
are outlined in Figure 6. The processes differ mainly in the
methods used to insolubilize the major proteins while the low
molecular weight components are removed.

The acid leach process begins by grinding defatted soy
flour or soy flakes to obtain a particle size of 95% through a
200 mesh U.S. Standard Sieve. Fine particles aid in extraction of the solubles and allow the finished product to be at-
omized for spray-drying. The flakes are extracted with water
which has been adjusted to the isoelectric point (pH 4.5) of
the soy protein with a food grade acid. This technique immediately immobilizes the protein and allows only the soluble
sugars and albumin to be leached during extraction. Extraction or leaching is carried out continuously for a specific
period of time, which is a function of the particle size, temperature and the agitation. Separation of the soluble soy
whey from the insoluble protein and cellulosic material is
carried out by centrifugation, dilution and a final concentration. The pH of insoluble polysaccharide-protein mixture is
raised to near neutrality and spray-dried (Moshy, 1964; Sair,
1959).

In the aqueous alcohol leach process, the nonprotein constituents are extracted with the appropriate mixture of alcohol and water, leaving the major protein fraction and the

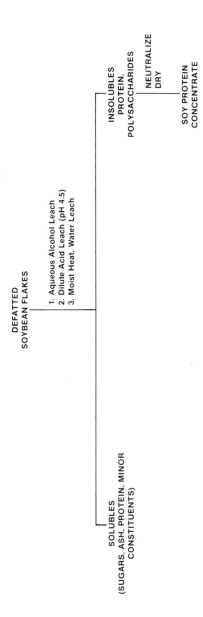

FIGURE 6.   Flow diagram of process for preparing soy protein concentrates.

polysaccharides, which are desolventized and dried (Mustakas, 1962 and O'Hara, 1965).

The moist heat water leach process takes advantage of the heat sensitivity of the soy proteins. The flakes or flour are heated to insolubilize the protein. The water soluble fraction containing primarily low-molecular weight constituents is extracted. The insoluble polysaccharides and major protein fractions are dried, yielding soy protein concentrate.

The typical composition of soy protein concentrates prepared by the three processes is given in Table 9. The acid leach process results in a product with high protein solubility, whereas the alcohol leach and the moist heat, water leach processes result in products with low protein solubility.

Soy protein concentrates can be texturized by processes similar to those described earlier for soy flour, to produce a range of products of various sizes, shapes, colors, and flavors, depending on the intended food application.

The insoluble polysaccharide fraction of the dehulled, defatted soybean meal is shown in Table 10. This fraction is primarily composed of hemicellulose and cellulose.

Isolated Soy Protein

Isolated soy protein is defined as the major proteinaceous fraction of soybean prepared from high quality, sound, cleaned, dehulled soybeans by removing a preponderance of the nonprotein components that shall contain not less than 90% protein (N x 6.25) on a moisture-free basis. The process for the production of isolated soy protein is outlined in Figure 7.

The usual starting material for isolated soy protein production is defatted soy flakes or flour which has high protein dispersibility. The extraction process involves wetting the soy flakes, with a proper amount of water, controlled temperature, and mixing the necessary amounts of high quality food grade chemicals for the defined length of time. The pH is closely defined or controlled throughout this step in the process. This is critical to the overall yield. After the protein has been solubilized, it is separated from the insoluble polysaccharides and crude fiber by centrifugation. The protein extract contains the soluble carbohydrates and the major protein fractions. Food grade acid is added to adjust the pH of the extract to approximately 4.5, resulting in precipitation of the major protein fractions. The precipitated protein is commonly referred to as soy protein curd. Washing of the soy protein curd is essential to remove undesirable carbohydrates such as a raffinose and stachyose along with color and flavor components.

The soy protein curd can be spray-dried, producing an isoelectric-type isolated soy protein product.

Table 9. Typical Composition of Soy Protein Concentrates

	Alcohol Leach	Acid Leach	Moist Heat Water Leach
Protein (N x 6.25)%	66	67	70
Moisture %	6.7	5.2	3.1
Fat %	0.3	0.3	1.2
Crude Fiber %	3.5	3.4	4.4
Ash %	5.6	4.8	3.7
Nitrogen Solubility Index %	5	69	3
pH (1:10 Water Dispersion) %	6.9	6.6	6.9

(Meyer, 1967)

Table 10. Insoluble Carbohydrates of Dehulled, Defatted Soybean Meal

Hemicellulose

Cellulose

Lignin

Pectin

Other Complex Carbohydrates

The pH of the soy protein curd can be adjusted with various cations to obtain a number of isolated soy protein products. Various mechanical and physical means are employed to prepare the neutralized soy protein curd for spray drying. The dried isolated soy protein is packaged and stored. The composition of a typical isolated soy protein is shown in Table 11.

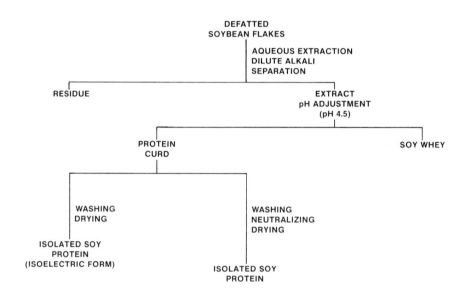

FIGURE 7. Outline of process for the production of isolated soy protein.

Table 11. Typical Composition of Isolated Soy Protein[1]

	%
Protein	92
Oil	0.5
Ash	4.5
Carbohydrates	0.3

[1]Moisture-Free Basis

The extractability of the proteins of defatted soybean flakes (or flour) as a function of pH, is shown in Figure 8 (Smith and Circle, 1938). This principle is employed for extraction and precipitation of major protein fractions of the soybean in the process for producing isolated soy protein. The major protein component of isolated soy protein is the globulin (Wolf, et al, 1962).

A wide variety of isolated soy protein products which have different functional properties are commercially available. The major functional properties are emulsification, fat absorption, water absorption, viscosity, gelation, fiber formation, and structure formation.

Isolated soy proteins can be texturized by several patented processes. Spun protein fibers, as described by Boyer

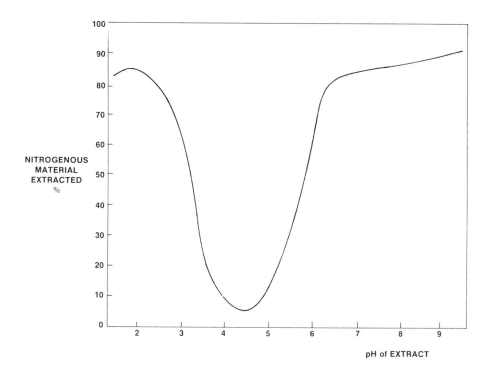

FIGURE 8. Effect of pH on the extractability of proteins from defatted soybean flakes (Smith and Circle, 1938).

(1954), are produced by passing an alkaline dispersion of soy protein through a spinnerette into an acid bath. The diameter of the fibers can range from 20 microns to 76 microns. This product is usually sold in a hydrated state.

Structured protein fibers are also commercially available and the process is described by Hoer (1972) and Frederiksen (1972). This product can be produced in a range of textural characteristics from tender to chewy, and is distributed in a hydrated (60-65% moisture) frozen state.

Amino Acid and Mineral Composition

The amino acid composition of soy protein products is important from a physical, chemical, and nutritional standpoint. The essential amino acid composition of typical soy flour, soy protein concentrate, and isolated soy protein is summarized in Table 12. The fractionation during processing accounts for differences in amino acid content between product categories. The sulfur containing amino acids, methionine and cystine, are considered to be the limiting amino acids in rat assays. The nutritional value of soy protein products will be discussed in detail in later chapters.

The mineral composition of soy flour, soy protein concentrates, and isolated soy protein is given in Table 13. These products contain nutritionally significant mineral such as calcium, iron, copper, phosphorus, zinc, and sodium.

FOOD USES OF SOY PROTEINS

Foods have unique chemical, physical and textural properties. The physical properties of meat, poultry, seafood, eggs, and dairy products are generally related to the proteins present in these products. The proteins in these foods contribute functional properties such as gelation, viscosity, emulsification, water absorption, dough formation, viscoelasticity, adhesion, cohesion, aeration, solubility, texture, flavor, and color.

The successful incorporation of soy proteins into traditional food products usually requires the protein ingredient to exhibit properties in the food product similar to that of the protein being supplemented, or replaced while not being detrimental to the overall quality. Soy proteins can also improve the characteristic properties of food products. The functionality properties of soy proteins are influenced by intrinsic variables of the food system such as pH, ionic concentration, solid content, processing conditions, etc. These

Table 13. Typical Mineral Content of Soy Protein Products

Element	Defatted[1] Soy Flour	Soy Protein[2] Concentrate	Isolated[3] Soy Protein
Arsenic	0.1 ppm	--	0.2 ppm
Cadmium	0.25 ppm[3]	--	<0.2 ppm
Calcium	0.22%	0.22%	0.18%
Chlorine	0.132%	0.11%	0.13%
Chromium	0.9 ppm[3]	<1.5 ppm	<1.0 ppm
Cobalt	0.5 ppm	--	<1.0 ppm
Copper	23 ppm	16 ppm	12 ppm
Fluorine	1.4 ppm	--	<10 ppm
Iodine	0.01 ppm	0.17 ppm	<10 ppm
Iron	110 ppm	100 ppm	160 ppm
Lead	0.2 ppm	--	<0.2 ppm
Magnesium	0.31%	0.25%	380 ppm
Manganese	28 ppm	30 ppm	17 ppm
Mercury	0.05 ppm[3]	--	<0.5 ppm
Molybdenum	2.6 ppm[3]	4.5 ppm	<3.0 ppm
Phosphorus	0.68%	0.70%	0.76%
Potassium	2.37%	2.1%	960 ppm
Selenium	0.6 ppm	--	0.36 ppm
Sodium	254 ppm	50 ppm	1.1%
Sulfur	0.25%	0.42%	--
Zinc	61 ppm	46 ppm	40 ppm

[1]*Kellor, R. L., J. Am. Oil Chem. Soc., 51, 77A, 1974.*
[2]*Anon., Technical Service Bulletin, Product Information Sheet: Promosoy-100, Central Soya, Chicago, Ill.*
[3]*Anon., Ralston Purina Company, 1978, Unpublished Data.*

Meat Products

One of the major uses of isolated soy protein is in comminuted or emulsified meat products, such as frankfurters, bologna and various loaves and coarse ground meat products.
Isolated soy proteins are being used as complementary proteins
to the meat proteins not only because of their moisture-
binding, fat emulsifying and emulsion stabilizing properties
in meat products, but, also, their nutritional and flavor
characteristics. According to Schweiger (1974) "Soluble soy
protein isolates are used mainly for their emulsifying capability, their emulsion stabilizing effect, and their property
of increasing viscosity and forming gels on heating." Isolat-

Table 12.   Essential Amino Acid Composition of Soy Derived
Protein Products

Amino Acid	gm Amino Acid/100 gm Protein		
	Soy Flour[1]	Soy Protein Concentrate[2]	Isolated Soy Protein[3]
Isoleucine	4.7	4.8	4.9
Leucine	7.9	7.8	7.8
Lysine	6.3	6.3	6.4
Methionine	1.4	1.4	1.3
Cystine	1.6	1.6	1.3
Phenylalanine	5.3	5.2	5.4
Tyrosine	3.8	3.9	4.3
Threonine	3.9	4.2	3.6
Tryptophan	1.3	1.5	1.4
Valine	5.1	4.9	4.7

[1]*Kellor, R. L., J. Am. Oil Chem. Soc., 51, 77A, 1974.*
[2]*Anon., Technical Service Bulletin, Product Information Sheet: Promosoy-100, Central Soya, Chicago, Ill.*
[3]*Anon., Ralston Purina Company, 1978, Unpublished Data.*

variables should be incorporated into laboratory methods designed for measuring the functional characteristics of soy proteins so that relationships may be established between basic properties and the food application. However, because of the complexity of the interaction between ingredients, it is preferred to evaluate the proteins in the actual food system to determine their applicability.

Major applications of soy protein products are in processed meat and fish products, bakery products, dairy-type products, infant formulas, protein supplements, hospital feeding, meat analog products, and a variety of formulated or fabricated food products.

ed soy protein is used in comminuted meat products to supplement the myosin and actomyosin in emulsifying and encapsulating fat to prevent fat separation and to bind water and especially meat juices during cooking. The above properties of isolated soy proteins allow for the high replacement of meat proteins in emulsified meat products. The replacement of high levels of meat proteins (30%-75%) requires soy products that are capable of imparting a characteristic meat-like texture (Roberts, 1974). At high levels of replacement the soy protein product must have the functionality characteristics to emulsify fat and water and contribute a physically desirable structure to the meat product, and also retain a good nutritional and flavor profile. For example, in frankfurters and bologna-type products, the degree of resiliency and firmness during mastication are two properties which contribute to the property commonly know as "bite" (Roberts, 1974). Isolated soy proteins have the ability to contribute these organoleptic properties to emulsified meat products.

Isolated soy proteins have multiple functional properties which contribute in different ways to an emulsified meat product. They are low in flavor and odor and contribute gel-like properties to an emulsified meat product in a manner similar to that of meat proteins (myosin and actinomysin). The other soy products usually have lower solubility characteristics and are used primarily for fat and water absorption properties. Isolated soy proteins have improved flavor and odor characteristics when compared to other soy products. This becomes more important in frankfurter-type products that are being marketed in the U.S. in which up to 40% of the protein present in these products is provided by isolated soy protein. The use of soy protein products allows meat emulsions to be prepared in a wider-range of emulsion temperatures than is possible when using meat proteins only. The use of isolated soy protein may permit a manufacturer to utilize a wider variety of meat cuts while maintaining overall product quality.

Textured soy flour and concentrates are used extensively in coarse, ground meat products and convenience food items, such as frozen dinners or frozen entree items that are served at home and in institutional food service, as extenders and for the maintenance of textural properties. The textured soy products can also contribute some water and fat absorption properties to these products in addition to their chewy and other textural characteristics. They are also used in pizza toppings, chilli products, meatballs, meat patties, tacos, meat spreads, poultry products and fish patties. Isolated soy proteins are used in combination with textured soy flour and concentrate products to provide binding, adhesive and cohesive properties.

The utilization of soy proteins in extending intact muscle tissue to increase finished product yield while maintaining protein equivalency in traditional products is described by Hawley, et al (1976). This development deals with the injection or pumping of ham or other intact muscle tissues with isolated soy proteins and brine solutions. A brine is prepared with isolated soy protein, salt, sugar, and flavorings. This brine mixture is injected into the muscle tissue and then the muscle tissue is massaged or tumbled to assist in the penetration and uniform distribution of the isolated soy protein in the muscle tissue. The meat is then stuffed in a casing, smoked and cooked according to the normal process. High quality hams with yields of 135% of green weight can be prepared. This procedure is a significant advancement since it is not necessary to grind, mince, or chop meat in order to extend it with soy protein products.

New meat products employing isolated soy protein are now entering the marketplace. These products utilize the lean meat portions of the animal in combination with fatty tissue and isolated soy proteins to yield a fabricated bacon-type product. The manufacturers are taking advantage of the binding properties of isolated soy proteins to produce a bacon-type product with decreased dimensional shrinkage and cooking loss during frying or cooking (Moore, 1978).

Isolated soy proteins are used in several meat products in Japan. The isolated soy proteins are used as binders in sectioned and formed hams, as stabilizers of emulsions in fish sausage, as binders in sausage for water retention, and as protein supplements. The soy protein is also being used in extending some of the traditional Japanese foods such as fish sausages, kamaboko, chikuwa, agekama, and pressed ham. The protein must contribute a gel-like texture with viscoelastic properties similar to that obtained with fish proteins.

Poultry rolls are being prepared with structured isolated soy proteins and powdered isolated soy proteins. These two products can complement each other with one providing the textural properties for the poultry roll and the other contributing to the binding or adhesion characteristics by forming a gel. The isolated soy protein in combination with structured soy products is especially effective in upgrading mechanically deboned meats to food products with acceptable form, color, flavor, and mouth feel.

Baked Products

Significant quantities of full-fat and defatted soy flour and soy concentrates have been used in the baking industry for many years. Recently isolated soy proteins are being used in these applications. Two general categories of full-fat and defatted soy flours are used, the enzyme active and enzyme in-active soy flours. Only a limited amount of full-fat soy flours is used in the United States. A much higher level of full-fat soy flour is being used in the United Kingdom and Europe. The functional properties, such as enzymatic activity and water absorption capacity, are important to some bakery systems. Enzyme active soy flours are used to improve the crumb color of bread and other yeast raised products, especially those which are not chemically bleached. Low levels of the full-fat soy flour added to wheat flour at levels of 1/2-1% of the wheat flour allows the production of bread with increased crumb softness and improved keeping quality (Pringle, 1974).

Defatted soy flour is used in combination with dried cheese whey for the replacement of non-fat dry milk in the production of white bread. The toasted soy flours are used to retain moisture and produce a softer, longer lasting product. The full-fat flours act as natural emulsifiers and stabilizers in baked goods as well as producing a product with a softer, moister crumb.

Soy flour at levels of 6-12% of the wheat flour has been used to produce high protein specialty breads (Hoover, 1974). Usually sodium stearoyl-2-lactylate or ethoxylated monoglycerides are utilized with the soy flour to maintain loaf size and structure. Tsen and Hoover (1973), and Pomeranz, et al (1969) demonstrated that bread with acceptable volume, grain and texture could be prepared with up to 24 lbs. of soy flour per 100 lbs. of wheat flour. This is a relatively economical way of providing an increased level of protein to the world population.

Soy flours are used at varying levels in doughnut and cake mixes. The soy flour helps regulate the amount of oil absorbed during the frying stage in doughnut production. The use of soy flours at a level of approximately 2% in commercial cake formulations appears to aid in producing cakes that are more tolerant to process and ingredient variations according to Cotton (1974).

Isolated soy proteins are being used in cake and doughnut mixes as replacements of non-fat dry milk. The isolated soy protein, when combined with corn syrup solids, dried cheese whey or other carbohydrates, functions as a replacement for non-fat dry milk in many of these applications.

Soy flours are used in some varieties of crackers. It is estimated by Cotton (1974) that the soy flour content of these specialty crackers is 2-5% of the total ingredient weight.

Tsen and Hoover (1973) used soy flour to develop a high protein cookie. Presently, a high protein cookie that uti-

lizes isolated soy proteins as the protein source is being
manufactured in Canada. Soy flour added to cookie doughs improves the release properties of the dough in mechanical operations, imparts a nutty flavor to the cookie and aids in the
emulsification of fats and other ingredients in the dough
(Levinson and Lemancik, 1974).

Infant Formulas and Food

Milk-free diets have been developed for use in the care of
infants with special nutritional requirements or that need
special care. The first formulas developed during the early
1950's utilized soy flour as the source of protein. These
were considered to be among the better milk substitutes available to provide the complete nutrition for the infant during
its early days of life. However, infants fed these formulas
often produced loose malodorous stools and this causes chafing
in the diaper area (Fomon and Filer, 1974). This condition
probably results from the presence of indigestible sugars. A
second generation of infant formulas based on isolated soy
proteins was developed during the early and mid 1960's. These
formulations had improved color, odor, flavor, and seldom
caused loose or malodorous stools. The indigestible carbohydrates or other irritating components are removed during the
production of isolated soy proteins (Fomon and Filer, 1974).
It has been estimated that in 1973 about 10% of the infants in
the U.S. were fed formulas based on isolated soy protein
(Fomon and Filer, 1974). The usage level is reported to be
increasing each year with formulas available in both liquid
(concentrated and ready-to-use) and dry form.

In addition to the milk-free infant formulas, special formulas utilizing isolated soy proteins have been developed for
older infants, geriatric, hospital, and postoperative feeding.
Low carbohydrate diets designed to aid in the diagnosis of di-
saccharidase deficiency or monosaccharide intolerance have
been developed. Soy flour, soy protein concentrates and isolated soy protein can be used to increase the protein content
of cereal products such as rice and wheat, when the infant begins to utilize solid foods.

Food Analogs

Several meat analog products are being produced that utilize soy flour, soy concentrate and isolated proteins. These
analog products usually contain other sources of protein such
as wheat gluten, egg albumin, and yeast. The soy proteins are
used in both the textured and powder forms. The production of
meat analog products and processing of textured products is

described by Rosenfield and Hartman (1974) and Horan (1974).

A variety of analog products are available for vegetarians or people who have special dietary needs. Several products have been developed and are marketed for breakfast, lunch and dinner meals. Various meat analog products are manufactured and marketed for use as seasonings for garnishes and related dishes.

Dairy Type Foods

Soy proteins are used in dairy-type products. In addition to the use of soy protein products as replacements of non-fat dry milk in baked goods and the use of isolated soy protein in infant formulas, isolated soy proteins are also being used in critically emulsified products, such as non-dairy coffee whiteners and whipped toppings. The functional properties are especially important in these products and appear to be more critical than for their application in meat and bakery products (Claus, 1974). The functional requirement for protein in these applications requires the refinement found only in isolated soy protein.

Properties of specific isolated soy proteins allow the protein to function in these applications in an excellent manner. Considerably more stress is placed upon a dry coffee whitener than a liquid coffee whitener. The use of isolated soy protein to replace approximately 50% of the sodium caseinate in spray dried coffee whiteners is described by Cho and Kolar (1977).

Isolated soy proteins are being used in liquid, aerosol and frozen-type whipped topping products. It is being used in evaporated-type milk products. Isolated soy protein is used as a protein source in a dry product designed for addition to milk by the consumer with the resulting product supplying the nutritional needs of a complete breakfast. A considerable amount of development activity is underway in the utilization of isolated soy protein in yogurt, sour cream, frozen desserts, cheese and dip-type products.

Isolated soy proteins and soy protein concentrates are used as protein sources in milk replacers for baby animals, such as pigs, lamb, and calves. In these formulations, the primary function of the protein is to serve as a protein source to provide nutrition to these animals.

Protein Supplements

Isolated soy protein, soy concentrates and soy flours are used as protein sources in the protein supplement industry, with isolated soy protein being the protein source of choice

by many of these manufacturers. Several companies manufacture
products that are 80% protein or greater and contain other
protein sources, vitamins and minerals, and flavorings. These
products are marketed as protein supplements to the normal diet.

Isolated soy protein and skim milk powders are the most
suitable sources of protein for slimming food products.
(Kolb, 1974.) The selection of isolated soy proteins is based
on its acceptable sensory qualities, favorable amino acid composition, high protein content with low non-protein calories
and with a relatively light color. The isolated soy protein
has good dispersibility, suspension properties and good storage stability.

Other Uses

Special soy protein products that are highly water soluble
are prepared by an enzymatic modification. These products
have the ability to form a foam when dissolved in water and
whipped, and are used for the aeration of food products. The
incorporation of air into a food product frequently will bring
about improvements in texture and consistency. These protein
based whipping agents are used in confectionery products,
marshmallows, biscuits, cookies, foam heads on soft drinks,
and prepared cake mixes, etc. (Mansvelt, 1974). The hydrolyzed soy protein whips faster and is more tolerant to over-
beating and higher temperatures than egg albumin. It has good
moisture retention which improves the shelf-life of marshmallows (Levinson and Lemancik, 1974). These soy products are
used in angel food box cake mixes and provide a foaming effect
in dry cocktail mixes.

Isolated soy protein and soy concentrate are being formulated in meal replacement products that are designed for
breakfast and snacks. Soy protein products are used in breakfast cereals to increase the protein content and to assist in
providing an overall complete nutritional breakfast.

Products manufactured from the whole bean, without defatting are being produced and are called soy nuts. These products are eaten directly as snacks and also are used in the
baking industry as a replacement of the more expensive nuts.

Soy flour or isolated soy proteins have been approved for
use in margarine as a part of the standards of identification
for margarine products; however, the amount of soy proteins
being used in this application is not known (Code of Federal
Regulations, 1977).

Soy protein food products play a major role in U.S., over-
seas and domestic food assistance programs (Senti, 1974).
Through these food assistance programs, the U.S. Agency for

International Development has donated food to many countries. Soy protein products were designed primarily for children in the developing nations. Textured soy protein products and protein fortified enriched macaroni were introduced to the U.S. school lunch program to meet part of the meat requirements in a Type Two school lunch program. In addition, a school lunch-breakfast program allowed the use of soy protein as an ingredient in protein fortified foods such as donuts, cake-like baked products and cereal-fruit products (Senti, 1974). Some of the fortified formulated food products used for distribution in foreign food programs are corn-soy blend (CSB) and corn-soy-milk mix (CSM).

ACKNOWLEDGMENT

The authors are grateful for the assistance of Wayne Moore, Ralston Purina Company, for his technical assistance in the manuscript preparation, tables and figures.

REFERENCES

Atkinson, W. T. (1978). U.S. Patent No. 3,488,770, assigned to Archer Daniels, Midland Company.

Boyer, Robert A. (1954). U.S. Patent No. 2,682,466, assigned to Ralston Purina Company.

Central Soya, Technical Service Bulletin, Product Information Sheet: *Promosoy*-100.

Cho, Iue, C. and Kolar, C. W. (1977). U.S. Patent No. 4,025,659, assigned to Ralston Purina Company.

Claus, W. S. (1974). *J. Amer. Oil Chem. Soc.*, 51, 197A.

Code of Federal Regulations, Office of Federal Register (1977), Title 21, Part 166, Subpart B, p. 330.

Cotton, Robert H. (1974). *J. Amer. Oil Chem. Soc.*, 51, 116A.

Flier, R. J. (1976). U.S. Patent No. 3,940,495, assigned to Ralston Purina Company.

Fomon, S. J. and Filer, L. J., Jr. (1974). *Infant Nutrition*, Second Edition, Ch. 15 "Milks and Formulas", p. 359.

Frederiksen, C. W. (1972). U.S. Patent No. 3,662,671, assigned to Ralston Purina Company.

Hawley, R. L., et al. (1976). U.S. Patent No. 3,989,851, assigned to Ralston Purina Company.

Hoer, R. A. (1972). U.S. Patent No. 3,662,672, assigned to Ralston Purina Company.

Hoover, William J. (1974). *J. Amer. Oil Chem Soc.*, 51, 186A.

Horan, Frank E. (January, 1974). *J. Amer. Oil Chem. Soc.*, 51, 67A, "Soy Protein Products and Their Production", p. 67A.

Kawamura, S. (1967). *Tech. Bull. Faculty Agric.*, Kagawa Univ., 18, 117.

Kellor, Richard L. (1974). *J. Amer. Oil Chem. Soc.*, 51, 77A.

Kolb, Erich (1974). *J. Amer. Oil Chem Soc.*, 51, 200A.

Levinson, A. A. and Lemancik, J. F. (January, 1974). *J. Amer. Oil Chem. Soc.*, 51, 135A

Mansvelt, J. W. (1974). *J. Amer. Oil Chem. Soc.*, 51, 202A.

Meyer, E. W. (1967). *Proc. Int. Conf. on Soybean Protein Foods*, U.S. Dept. Agr., ARS 77035, p. 142.

Moore, Karen (1978). *Food Product Development*, 12, 98 and 101.

Moshy, R. L. (1964). U.S. Patent No. 3,126,286, assigned to General Foods Corp.

Mustakas, G. C., et al (1962). *J. Amer. Oil Chem. Soc.*, 39, 222.

Mustakas, G. C., et al (1970). *Food Technol*, 24, 1290.

Official Grain Standards of the United States Grain Division (1977). Consumer and Marketing Service, U.S. Dept. Agr., Hyattsville, MD.

O'Hara, J. B. and Schoepfer, A. E. (1965). U. S. Patent No. 3,207,744, assigned to A. E. Staley Mfg. Co.

Pomeranz, Y., M.D., et al (1969). *Cereal Chem.*, 46, 503 and 512.

Pringle, W. (1974). *J. Amer. Oil Chem. Soc.*, 51, 74A.

Rackis, J. J., et al (1970). *J. Agr. Food Chem.*, 18, 977.

Ralston Purina (1978). Unpublished Data.

Roberts, Leonard H. (1974). *J. Amer. Oil Chem. Soc.*, 51, 195A.

Rosenfield, D. and Hartman, W. E. (1974). *J. Amer. Oil Chem. Soc.*, 51, 91A.

Sair, Lo (1959). U.S. Patent No. 2,881,076, assigned to Griffith Laboratories, Inc.

Schweiger, R. G. (1974). *J. Amer. Oil Chem. Soc.*, 51, 192A.

Senti, F. R. (1974). *J. Amer. Oil Chem. Soc.*, 51, 138A.

Smith, A. K. and Circle, S. J. (1938). *Ind. Eng. Chem.*, 30, 1414-1418.

Smith, A. K. and Circle, S. J. (1972). Soybeans: Chemistry and Technology, Vol. I, Ch. 9, p. 311.

Strommer, P. K., and Beck, C. I. (1973). U.S. Patent No. 3,754,926, assigned to General Mills.

Tombs, M. P. (1967). *Plant Physiol.*, 42, 797.

Tsen, C. C. and Hoover, J. J. (1973). *Cereal Chem.*, 50, 7.

Wilding, M. D. (1974). *J. Amer. Oil Chem. Soc.*, 51, 128A.

Wolf, W. J., et al (1962). *Biochem. Biophys.*, 99, 265.

Wolf, W. J. and Cowan, J. C. (1975). *Soybeans as a Food Source*, Revised Edition, "Seed Structure and Composition", p. 2.

Wolford, Kenneth M. (1974). *J. Amer. Oil Chem. Soc.*, 51, 131A.

DISCUSSION

Altschul: The various fractionations that you have reported on, do they reflect the information, let's say, on the cellular location of the various fractions of the seed and on the information that is coming out about the various protein fractions, and do you see that perhaps there could be newer types of fractionations evolving out of this knowledge?

Waggle: Aaron, that is a loaded question. From the standpoint of processing into isolates, we are recovering part of the 2S, most of the 7S, 11S and 15S fractions. The 7S and 11S fractions are the major proteins present and they are likely to be the primary contributors to functionality of isolates. Of course, soy flours contain all of the protein fractions found in soybeans. In soy protein concentrates made by the acid leach process, the low molecular weight proteins are partially removed. I do forsee that the major protein fractions could be separated to make commercial products. I think this will depend on the application in the sense that the protein will have to perform a particular function in the food system.

THE IMPORTANCE OF FUNCTIONALITY
OF VEGETABLE PROTEIN IN FOODS

Wilda H. Martinez

Functionality has been defined as any property, other than nutritive value, that affects the utilization of a product (Pour El, 1976). This is a very broad definition that logically and easily could be expanded to include nutrient quality. The key words in this definition are "property...that affects utilization..." For example, if the expanded definition were used, nutritive quality could be defined as the primary and essentially sole functional characteristic affecting the use of soybean flour in cereal-based mixtures such as CSM (corn, soy, milk blends) and composite flours. Whereas, for other end uses, flavor might be identified as a limiting or negative functional property. Thus, under such a definition, functionality could be positive or negative, would be end use dependent, and would include flavor, color and aroma characteristics of a protein ingredient in addition to the numerous properties usually listed, such as water adsorption, fat adsorption, gelation, emulsification, viscosity, etc.

A definition as universal and pliant as this suggests that it was derived from a marketing perspective primarily to identify and catalogue reasons for product acceptance or rejection. Unquestionably, it is important in the marketing of a product to be cognizant of the properties that determine acceptability as a food or food ingredient. If one is interested in nutritional intervention and dietary enrichment, acceptability might be considered essential. However, to list all properties positive and negative under a single collective term hinders rather than helps in understanding the determinants of these properties, which are the key to increased acceptance.

Three major factors govern food product acceptance. These are color, flavor including aroma, and texture. Each of these factors appeals to a different set of the senses and varies in relative importance with the type of product. With requisite

ISBN 0-12-751450-3

information and appropriate education of the populace, a
fourth factor, nutritive value, could be added. Each of these
factors is determined by a set of properties of the food or
food ingredient. From this perspective, functionality is immediately divided into four categories and limited to the positive aspects of acceptance. Functionality can be defined,
therefore, as the set of properties that contributes to the
desired color, flavor, texture or nutritive value of a product. To prevent the loss of an excellent term through indis-
criminate overuse, I would like to propose that in future discussions the term functionality be specified by coupling with
it an appropriate modifier, i.e., color, flavor, texture or
nutritive.

In the use of vegetable proteins as food ingredients,
there are very few examples of color or flavor functionality.
Undenatured soy flour is used as a bleaching agent for wheat
flour, a property attributed to the lipoxygenase content of
the soy (Wood, 1967). Most defatted oilseed flours--soybean,
cottonseed and peanut, provide improved crumb color in baked
goods, presumably due to increased browning reaction between
the oilseed proteins and carbohydrates (Turro and Sipos, 1968;
Kahn, et al, 1975, 1976). Also the pigments of cottonseed
flour are reported to enhance the yellow color of doughnuts
(Olson, 1972).

The flavor characteristics of vegetable protein products
usually impart undesirable rather than functional properties.
Cottonseed products are said to produce a desirable nut-like
flavor in certain baked goods. However, in the development of
food products, change in flavor, be it positive or negative,
is usually considered undesirable with regard to acceptance.
Indeed, the major impetus for producing the 70 percent soy
concentrate described by Dr. Waggle (this conference) was to
modify the characteristic soy flavor and produce a bland product. Color and flavor, therefore, are not the primary functional properties motivating the use of vegetable protein
products in food systems.

The primary functionalities for which vegetable protein
products are utilized are their contribution to the textural
and nutritive properties of a food. Since the nutritive value
of soy protein products is the principal subject of discussion
at this conference, I will limit my remarks to the textural
aspects of functionality.

In foods, texture is usually delineated by one or more
descriptive terms such as those listed in Table 1. Based on
the early definitions of Kramer (1959) and Matz (1962), texture can be described as the mingled experience derived by the
skin or muscle senses of the mouth upon ingestion of a food or
beverage exclusive of temperature and chemically initiated

Table 1. *Listing of Texture Descriptors*[1]

Hardness	Crispness
Viscosity	Creaminess
Adhesiveness	Spreadability
Cohesiveness	Toughness
Chewiness	Crunchiness
Guminess	Tenderness
Springiness	Juiceness
Mushiness	

[1]*Moskowitz and Kapsalis, 1976.*

taste. In 1970, Corey wrote texture "is but another name for
the interaction of the human with the mechanical properties of
the material" and Sherman (1970), in a modification of Szczesniak's (1963) early definition, described texture as "the composite of those properties which arise from the structural
elements and the manner in which it registers with the physiological senses."
  Szczesniak (1977a) in a recent overview on food texture
research points out that the latter definition "recognizes
three essential elements of texture:
1) that it is a sensory quality;
2) that it stems from the structural parameters of the food
(molecular, microscopic or macroscopic); and
3) it is a composite of several properties."
Among these properties are mechanical properties; visual properties, i.e., the "open" or "closed" structure of cakes, audio
properties (noise); and properties which are the result of
processes such as rate of size reduction, moisture uptake or
release, oil or fat release and melting.
  In the evaluation of texture, the description of the mechanical properties of foods in terms of basic rheological
phenomena initially has received the greatest attention. Rheology has been used to address three different aspects of the
problem of food texture: basic rheological structure of
foods, processing and quality, and consumer acceptance (Szczesniak, 1977b). By the application of stress and deformation
over time the rheological structure of a food can be described
mathematically and characterized in terms such as Newtonian,
plastic, pseudoplastic, elastic, retarded elastic, viscoelastic, and thixotropic (deMan, 1976). Rheological characterization of food systems, particularly purees and emulsions, is
critical to the processing and movement of these materials

through pipes, nozzles, pumps, etc., and to understanding heat transfer in evaporators, retorts and coolers. Rheological characterization of foods with respect to evaluation of food quality and consumer acceptance thus far has not proved fully satisfactory.

The area of food quality is replete with empirical methods for quantification of mechanical properties (Szczesniak, 1973). In addition to problems with calibration of the instrumentation, these methods suffer from a lack of standard calibration material and from the fact that most tests are "replicable," i.e., they can be repeated on fresh samples, but few are "reproducible," i.e., repeatable on the same sample (Scott Blair, 1976). "Replicable" tests tend to describe "processes," i.e., the penetration of the structure of a pea, rather than the mechanical properties or tenderness of a pea, and thus limit generalization of the methodology to other foods.

In the area of consumer acceptance, there arises the added problem of relating rheological measurements to sensory evaluation. With the realization that texture is a spectrum of properties, the principle of texture profiling was developed. Texture profiling is achieved by use of either a single test from which a number of properties can be characterized or by use of several tests, each detecting one or more properties that may be related to sensory texture terms (Breene, 1975; Szczesniak, 1975). However, this relationship is complicated by the inherent psychological factors of sensory evaluation (Scott Blair, 1976; Moskowitz and Kapsalis, 1976) and a lack of quantification of the stress/strain conditions occurring in the mouth during evaluation. To assist the understanding of sensory perception of rheological stimuli, a branch of psychology called psychorheology has been developed (Moskowitz, 1977). Recognition of the multifaceted nature of texture has also raised the question as to whether a rheological approach can provide a full description of texture (Bourne, 1975, 1977).

The complexity of the problem of texture evaluation and consumer acceptance increases inordinately when one attempts to use rheological, empirical or imitative model systems to predict the functional contribution of a food ingredient. Application of rheological principles to individual ingredients in model systems as a predictive measure of texture forming capability in a food system is a misuse of the tool and as a result the literature is replete with information of little or no value. Rheological assessment of mechanical properties provides a summary statement on the texture of a whole, intact food. Rheological measurement does not provide the ability to evaluate the relative contribution of the structural components of the food and the influence of physical and chemical

environment on the individual components of the intact food.
For example, in recent studies on the texture of spun soy fiber, Stanley, et al (1971) and Cumming, et al (1972) found
that compared to meat, the spun fibers were very uniform and
homogenous, exhibited higher breaking strength and greater
elongation, were less able to return to the original form upon
removal of stress and relaxed more under the applied stress.
This type of evaluation tells us a great deal about the difference between spun fibers and meat, but provides essentially
no indication about the means to improve the textural quality.   Improvement must be made on the basis of trial and error.
     Prediction of the manner in which a specific ingredient
will perform requires an adequate knowledge of:
1)   the physical structure of the ingredient,
2)   the chemical structure and resultant physical properties
of the constituents contained in the ingredient, and
3)   the effect of the physical and chemical environment of the
particular food system on the ingredient.
Inherent in this statement is an expression of need not only
for increased information on the nature of food ingredients
but also increased knowledge on the construct of the foods
which we are attempting to imitate or extend.
     A number of studies have provided information on the
structure of oilseeds and their processed products (Dieckert
and Dieckert, 1972; Engleman, 1966; Saio and Watanabe, 1968;
Wolf and Baker, 1972, 1975; Yatsu, 1965).   The mature cell of
an oilseed is a very highly organized, densely packed structure in which nature has neatly stored nutrients for the development of the new plant. Lipids are deposited in the spherosomes of the cell and proteins concentrated in the protein
bodies.   The scanning electron micrograph in Figure 1A provides a three-dimensional orientation to the cell of the soybean and the protein body inclusions.   Transmission electron
micrographs of stained and fixed sections of the desiccated
seed provide a more detailed view of the ultrastructure (Figures 2-6).   Protein bodies are surrounded by a network of the
smaller, lipid storage particles--spherosomes and in the cottonseed (Figure 3) there is a third particle located within
the protein body.   These particles, called globoids, were isolated by Lui and Altschul (1967) and shown to be the storage
site of the seed phosphorus deposited as insoluble phytates.
Soybeans differ from cottonseed, peanuts and other dicotyle-
dons.   Soybeans do not contain globoids.   The cytological location and dominant chemical form of the phytates of the soybean are not known.   The mature soybean, however, may contain
starch granules (Figure 4).

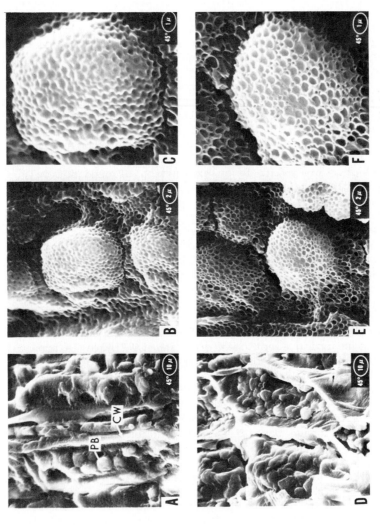

FIGURE 1. Scanning electron micrographs of freeze-fractured soybean cotyledon (A); protein body covered with sponge-like cytoplasmic network and spherosomes in fracture surface (B, C); hexane defatted fracture surface (D); protein body in defatted fracture surface showing cytoplasmic network (E, F). Courtesy of W. J. Wolf and F. L. Baker.

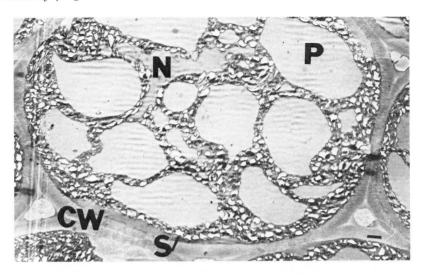

FIGURE 2. Transmission electron micrograph of a typical
spongy parenchymal cell from the cotyledon of a dry soybean;
N-nucleus, P-protein body, S-spherosome, CW-cell wall. Magnification 8000X - 1 micron. Courtesy of L. Y. Yatsu and T. J.
Jacks.

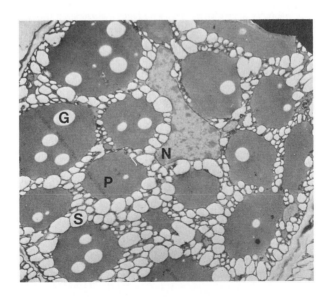

FIGURE 3. Transmission electron micrograph of a spongy
parenchymal cell from cotyledon of dry cottonseed; N-nucleus,
P-protein body, S-Spherosome, G-globoid. Taken at 7000X.
Courtesy of L. Y. Yatsu.

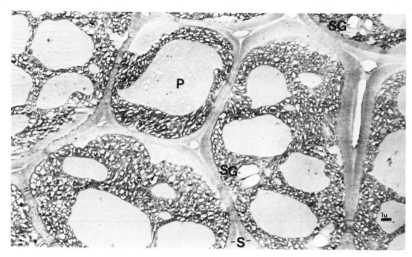

FIGURE 4. Transmission electron micrograph of several
spongy parenchymal cells from cotyledon of dry soybean; P-
protein body, S-spherosome, SG-starch granules. Magnification
6000X - 1 micron. Courtesy of L. Y. Yatsu and T. J. Jacks.

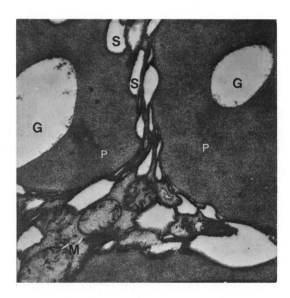

FIGURE 5. Transmission electron micrograph of a spongy
parenchymal cell from cotyledon of dry cottonseed; P-protein
body, S-spherosome, G-globoid, M-mitochondria. Taken at
24,000X. Courtesy of L. Y. Yatsu.

In addition to protein bodies and spherosomes, the cell
also contains the normal complement of structures essential to
cellular integrity embedded in and surrounded by the basic cytoplasm. Figure 5 illustrates the relative size of the various particles including the mitochondria (see also Figure 1A-
C). Protein bodies generally range in size from 5 to 20 microns while spherosomes are usually 1 to 3 microns in diameter. Detailed examination of these structures indicates that
the protein body is surrounded by a classical unit membrane
(Figure 6). The globoid lacks any identifiable membrane, and
the membrane surrounding the spherosome differs from that of
the protein body in structure (Yatsu and Jacks, 1972) and composition (Jacks, 1978).

Hydration of the seed, even soaking overnight, will reduce
the sharp convolutions in shape of these particles, particularly the spherosomes (compare Figure 2 and Figure 7), but
will not destroy the integrity of the subcellular structures.
Hexane extraction removes the neutral lipids of the sphero-

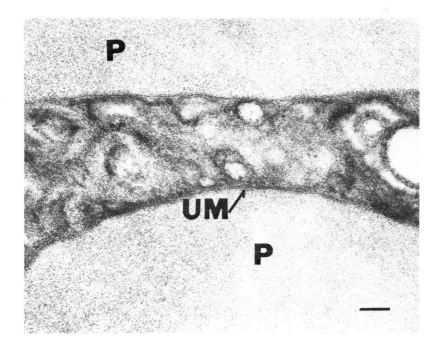

FIGURE 6. Transmission electron micrograph of a spongy
parenchymal cell from cotyledon of dry soybeans at high magnification; P-protein body, UM-unit membrane. Magnification
178,750X-0.1 micron. Courtesy L. Y. Yatsu and T. J. Jacks.

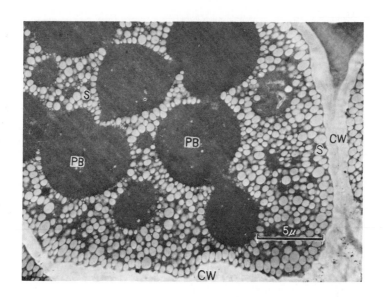

FIGURE 7. Transmission electron micrograph of soybean
cotyledon cells after soaking in water overnight; PB-protein
bodies, S-spherosomes, CW-cell wall.

somes (Figure 1 D-F). But the degree of spherosome membrane
disruption and coextraction will depend upon the moisture content of the seed, polarity of the extracting solvent, and mechanical work used in the extraction. The degree of cellular
disruption and protein body disruption is also dependent upon
the mechanical work (flaking, transport and grinding) and the
conditions used in the desolventizing-toasting operations.
Examination with the scanning electron micrograph (Figure
8A-C) suggests that some of the protein bodies remain intact
through all commercial operations used in the production of a
high NSI (nitrogen solubility index) soybean flour and even
through commercial alcohol-leaching for the production of a
soybean concentrate. However, in both the high NSI soy flour
and the alcohol-leached concentrate the distinctive cytoplasmic network surrounding the protein bodies has been converted
to an amorphous mass. Concentrates produced by heat denaturation and water washing or water extraction at acid pH lack any
recognizable structures.
    Based on this subcellular structure, the proteins of oil-
seeds can logically be classified into two groups--the storage
proteins, i.e., the proteins within the protein bodies, and
the nonstorage proteins, those proteins that are part of the
structural (membranes, cell wall) and functional (enzymes, nu-

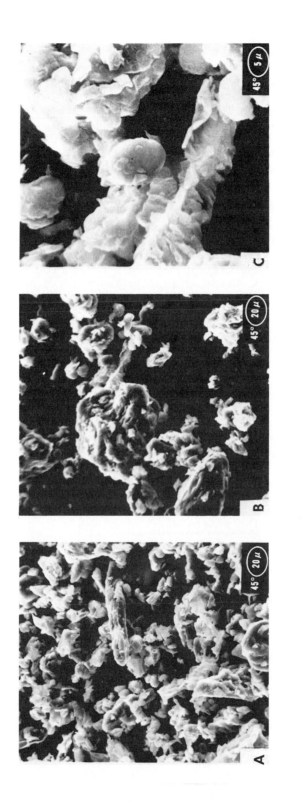

FIGURE 8. Scanning electron micrograph of high NSI commercial defatted flour at low magnification (A), and a commercial soy protein concentrate at low magnification (B) and at high magnification (C).

cleoproteins, cytoplasm) components of the cell. These structural differences also result in differences in physical chemical properties. Storage proteins, the major proteins in the
mature seed, are generally few in number but high in molecular
weight (greater than 100,000). The nonstorage proteins are
low in molecular weight and many in number. They are also
usually high in cystine and therefore capable of inter- and
intramolecular bonding.

By utilizing differences in the physical characteristics
such as aqueous dispersibility, the storage and nonstorage
proteins of the seed can be separated. With the cottonseed
this is accomplished by successive extraction first with water
and then alkali or salt (Berardi, et al, 1969; Martinez, et
al, 1970). Differences in molecular weight are readily apparent from the ultracentrifugal patterns of the protein isolates
(Figure 9). Because the undenatured storage proteins of the
soybean hydrate and disperse rapidly in water, 0.008 M calcium

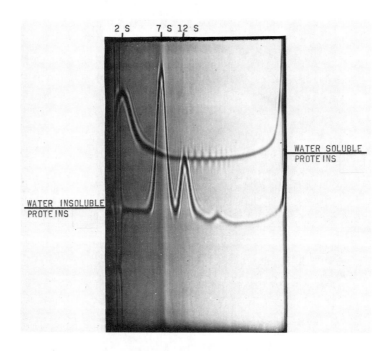

FIGURE 9. Ultracentrifuge patterns of isolate from successive water and alkaline extracts of defatted cottonseed
flour. Ratio 1g to 15ml. Upper pattern, water extract; Lower
pattern, alkaline extract. Buffer 0.2M Na_2CO_3-$NaHCO_3$,
pH 10.5. Analytical rotor. Taken at 59,780 rpm with standard
and wedge quartz cells. Sedimentation from left to right.

chloride must be used in the initial extraction to achieve a similar type of separation (Martinez and Berardi, 1971; Saio et al, 1973). Microscopic examination suggests that at low ionic strength the divalent calcium ion interacts with the protein body membrane to prevent hydration and dispersion of the storage proteins. Ultracentrifugal analysis (Figure 10) shows that in the soybean both groups of proteins contain a protein with a sedimentation coefficient of 7S, i.e., a molecular weight greater than 100,000.

In the cottonseed, the solubility characteristics of the two groups of proteins differ markedly (Figure 11). The storage proteins have a minimum solubility at pH7 with maxima at pH3 and pH10, while the nonstorage proteins have a minimum at pH4 and a maximum at pH7. A mixture of the two in proportion to their natural occurrence provides still a third solubility pattern. In the soybean, the solubility characteristics of the two groups of proteins are not as divergent.

The most important characteristics of the storage proteins with respect to functionality may be the dissociation--aggre-

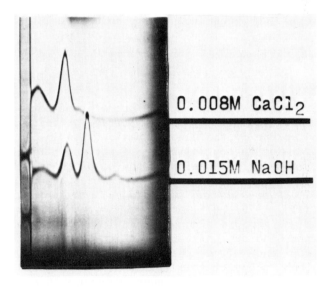

FIGURE 10. Ultracentrifuge patterns of isolates from successive dilute calcium chloride (0.008m) and alkaline extracts of defatted soybean flour. Ratio 1g to 15ml. Upper pattern, calcium chloride extract. Lower pattern, alkaline extract. Buffer K_2HPO_4-KH_2PO_4 pH7.6, 0.01M mercaptoethanol, ionic strength 0.5. Analytical rotor at 59,780 rpm with standard and wedge quartz cells. Sedimentation from left to right.

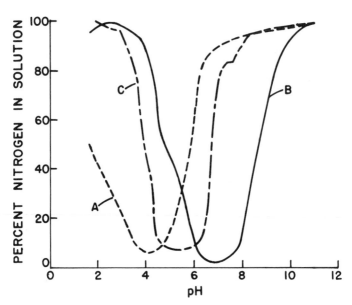

FIGURE 11. Nitrogen solubility curves of 1% solutions of
cottonseed isolates. Curve A - nonstorage protein isolate,
Curve B - storage protein isolate, Curve C - mixture of both
isolates.

gation phenomena exhibited by the proteins of both oilseeds.
The unique acid solubility of the storage protein isolate of
cottonseed is due to the electrostatic dissociation of the 7S
and 12S proteins to low molecular weight subunits (Figure
12). The 7S and 11S proteins of the soybeans are reported to
dissociate with changes in ionic strength, pH and in the presence of disulfide-cleaving agents such as cysteine, and sodium
sulfite (Wolf, 1970, 1976), whereas, aggregation is influenced
by temperature and protein concentration in addition to those
environmental factors influencing charge on the molecule.
    The importance of net charge to intermolecular reactions
and functional properties is illustrated in Figure 13 by the
variation in texture obtained upon heating a 10 percent aqueous dispersion of the cottonseed storage protein isolate at
different pH values (Berardi and Martinez, 1974). Such interactions are attributed to regions of both charge and hydropho-
bicity in the molecule. The relationship of functionality to
amino acid content, amino acid sequence, and protein size and
configuration is also illustrated by the fact that the presence of as little as one percent of the low molecular weight,
nonstorage proteins will prevent gelation and texturization of

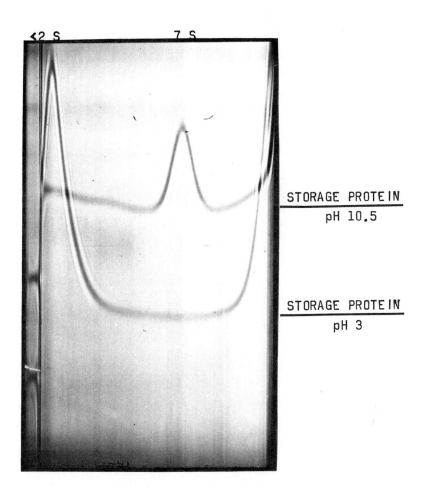

STORAGE PROTEIN
pH 10.5

STORAGE PROTEIN
pH 3

FIGURE 12. Ultracentrifuge pattern of storage protein isolate in acid and alkaline medium, analytical rotor at 59,780 rpm with standard and wedge quartz cells, 66 minutes post top speed, protein concentration 1.5%. Lower pattern - 0.2M citric acid, Upper pattern - 0.2M NA_2CO_3 - $NaHCO_3$, pH 10.5. Sedimentation from left to right.

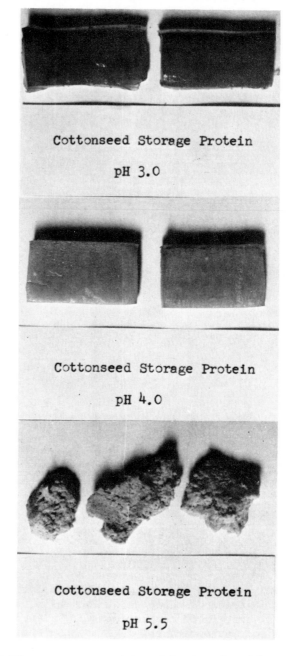

FIGURE 13. Textures achieved by heating 14 percent aqueous slurries of the storage protein isolate of cottonseed at
different pH values.

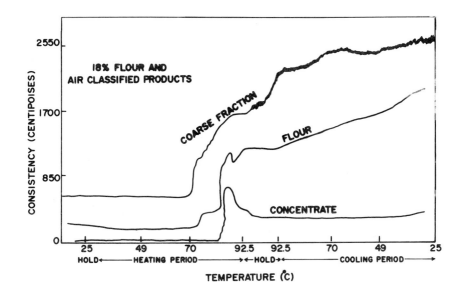

FIGURE 14. Consistency profiles of aqueous slurries containing 18 percent of either whole defatted glandless cottonseed flour or the air classified protein concentrate (70%) or coarse fraction from the flour.

the storage protein isolate. In a series of studies on the relative contribution of the 11S and 7S proteins of the soybean, Saio et al (1969, 1971, 1973, 1974, 1975) have shown that the 11S protein is primarily responsible for the formation and extensibility of tofu gel.

Wolf (1970) has noted that although functional properties are usually attributed to the proteins, other constituents of the defatted flours and concentrates can contribute to the functionality. This is particularly true for the cell wall constituents. It should also be recognized that unless the integrity of these seed structures are sufficiently disrupted by processing and/or the environment of the food system into which it is incorporated, the proteins will not contribute to the observed properties. Heat denaturation of the proteins during processing will also limit and possibly change the degree of functionality observed.

Because of the inherent aqueous insolubility of the proteins in the protein bodies of the cottonseed, cottonseed flours and concentrates provide excellent materials to illus-

trate the importance of the cellular components and heat to
the consistency of aqueous dispersion. If a defatted flour is
fractionated by air classification (Martinez, 1970) into a
fine fraction consisting predominantly of intact protein bodies, and a coarse fraction consisting predominantly of cell
wall fragments, and aqueous slurries of the two are compared
to a slurry of the original flour, one obtains the variation
in consistency with heating and cooling shown in Figure 14 as
measured with a Brabender Amylograph. It is evident that the
cell wall constituents are the major contributors to the consistency of these dispersions (Berardi and Martinez, 1970).
The importance of protein body rupture and solubilization of
the proteins is illustrated in Figure 15 by the use of pH to
increase solubility of the storage proteins in the defatted
flour. Note the very interesting thixotropic-like effect with
increase in temperature, and the minimal change in consistency
at pH 3.5 which is quite different from the change obtained

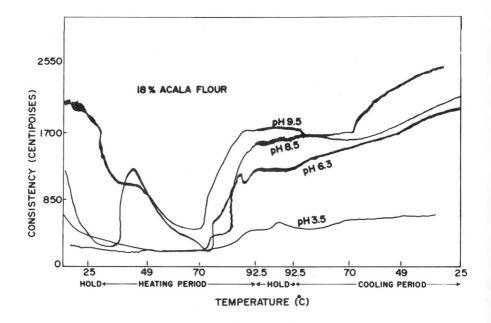

FIGURE 15. Consistency profiles of 18 percent aqueous
slurries of defatted glandless cottonseed flour at different
pH values.

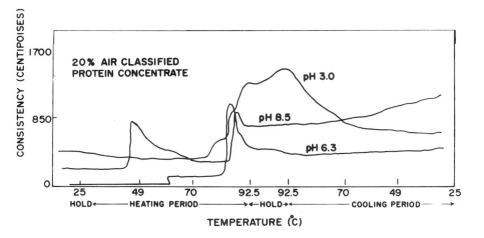

FIGURE 16. Consistency profiles of 20 percent aqueous slurries of air-classified glandless cottonseed protein concentrate (70%) at different pH values.

with the air classified protein body concentrate at pH 3.0 (Figure 16). At pH 6.3 there is a sharp increase in consistency of the concentrate slurry at about 87° with a subsequent rapid reduction on cooling. These changes suggest an initial aggregation followed by coagulation and separation into two phases rather than increased solubility of the proteins. The pH of the environment will also affect the consistency of the cell wall fragments in the air-classified coarse fraction (Figure 17).

Protein content alone is not sufficient to predict results. Knowledge of the type of protein, the quantity of protein, the nonprotein constituents and the environmental factors such as the presence of calcium ions are all necessary to understand the functional properties of a product. For example, the consistency patterns of three concentrates prepared by three different procedures—air classification, aqueous extraction and dilute calcium chloride extraction—are quite different (Figure 18). The two aqueous extracted concentrates contain the full complement of cell wall fragments present in the original defatted flour while the air classified concentrate contains very little. The difference between

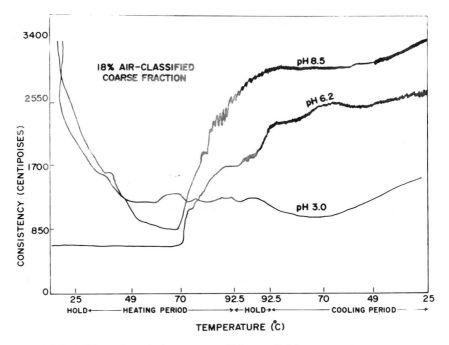

FIGURE 17. Consistency profiles of 18 percent aqueous
slurries of coarse fraction from an air-classified glandless
flour at different pH values.

curves B and C must be due to the interaction of the calcium
ion with the cell wall components.

At this point one might well ask why such an emphasis on
functionality at a meeting concerned primarily with the nutritional importance of vegetable proteins for humans. The response is quite simple and direct. If proteins are not provided in an acceptable form they will not be consumed. Humans
can be and are quite selective in the foods they eat. Unless
they are motivated by some medical or pseudo-medical reason,
they select foods that have desirable appearance, flavor and
texture solely for the pleasure of eating; not for nutritional
reasons. In essentially any example of a fabricated or processed food that may be cited such as mayonnaise, cakes, margarine, candies or ice cream, the eggs and milk are incorporated into the formulation not to provide nutrition, but rather
to provide pleasure - again primarily in the form of acceptable texture.

However, as was noted earlier, should the preliminary data
on the importance of vegetable proteins to the physiological

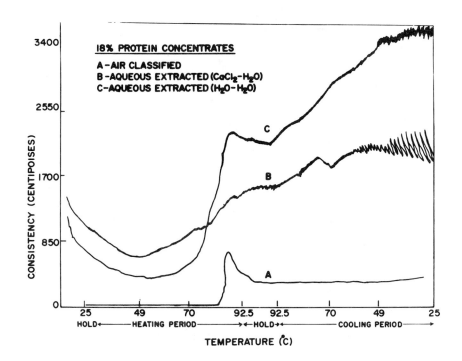

FIGURE 18. Consistency profiles of 18 percent slurries of cottonseed protein concentrates prepared by air classification or successive extraction with either water or dilute calcium chloride (0.008M) and water.

health of the population be proven, particularly with respect to heart disease, then we soon may be using the term nutritional functionality.

ACKNOWLEDGEMENT

I am indebted to Dr. K. Saio for permission to use Figure 7, to Dr. W. Wolf for supplying prints and permission to use Figures 1 and 8, and to Dr. L. Yatsu and Dr. Tom Jacks for supplying prints and permission to use Figures 2, 3, 4, 5 and 6.

REFERENCES

Berardi, L. C.; Martinez, W. H.; Fernandez, C. J. (1969). *Food Technol.* *23,* 75-82.

Berardi, L. C.; Martinez, W. H. (1970). Abstract 98 "IFT Program, 30th Annual Meeting, Institute of Food Technologists." May 24-28, 1970, San Francisco.

Berardi, L. C.; Martinez, W. H. (1974). Abstract 22 "IFT Program, 34th Annual Meeting, Institute of Food Technologists, May 12-15, 1974, New Orleans.

Bourne, M. C. (1975). *J. Text. Studies 6,* 259-262.

Bourne, M. C. (1977). *J. Text. Studies 8,* 219-227.

Breene, W. M. (1975). *J. Text. Studies 6,* 53-82.

Corey, H. (1970). *CRC Crit. Rev. Food Technol. 1,* 161-198.

Cumming, D. B.; Stanley, D. W.; deMan, J. M. (1972). *Can Inst. Food Sci. and Technol. J. 5,* 124-128.

deMan, J. M. (1976). "Rheology and Texture in Food Quality" (J. M. deMan, P. W. Voisey, V. F. Rasper, D. W. Stanley, eds.), pp. 8-27, AVI, Westport, Ct.

Dieckert, J. W. and Dieckert, M. C. (1972). "Symposium: Seed Proteins" (G. E. Inglett, ed.) pp. 52-85 AVI, Westport, CT.

Englemann, E. M. (1966). *Am. J. Bot. 58,* 231-237.

Jacks, T. J. (1978). Private Communication.

Kahn, M. N.; Rhee, K. C.; Rooney, L. W.; Cater, C. M. (1975). *J. Food Sci. 40,* 580-583.

Kahn, M. N.; Lawhon, J. T.; Rooney, L. W., Cater, C. M. (1976). *Cereal Chem. 53,* 388-396.

Kramer, A. (1959). *Food Technol. 13,* 733-736.

Lui, N. S. T., and Altschul, A. M. (1967). *Arch. Biochem. Biophys. 121,* 678-684.

Martinez, W. H.; Berardi, L. C.; Goldblatt, L. A. (1970). "Proc. Third Intern. Congr. Food Sci. Technol. (SOS/70)," pp. 248-261.

Martinez, W. H. and Berardi, L. C. (1971). U.S. Pat. 3,579,496 (U.S. Dept. of Agric.), May 18.

Matz, S. A. (1962). "Food Texture." AVI, Westport, CT.

Moskowitz, H. R. and Kapsalis, J. G. (1976). "Rheology and Texture in Food Quality" (J. M. deMan, P. W. Voisey, V. F. Rasper, D. W. Stanley, eds.), pp. 554-581, AVI, Westport, CT.

Moskowitz, H. R. (1977). *J. Text. Studies 8,* 229-246.

Olson, R. L. (1973). *Oil Mill Gaz. 77,* 7-8.

Pour El, A. (1976). "World Soybean Research" (L. Hill, ed.), pp. 918-946, The Interstate Printers & Publishers, Inc., Danville, IL.

Saio, K. and Watanabe, T. (1968). *J. Food Sci. Technol. (Tokyo) 15,* 290-296.

Saio, K.; Kamiya, M.; Watanabe, T. (1969). *Agr. Biol. Chem. 33,* 1301-1308.

Saio, K.; Kajikawa, M.; Watanabe, T. (1971). *Agr. Biol. Chem. 35,* 890-898.

Saio, K.; Watanabe, T.; Matsumi, K. (1973). *J. Food Sci. 38,* 1139-1144.

Saio, K.; Terashima, M.; Watanabe, T. (1974). *J. Food Sci. 40,* 537-540, 541-544.

Saio, K.; Sato, I.; Watanabe, T. (1975). *J. Food Sci. 39,* 777-782.

Scott Blair, G. W. (1976). "Rheology and Texture in Food Quality" (J. M. deMan, P. W. Voisey, V. F. Rasper, D. W. Stanley, eds.), pp. 1-7, AVI, Westport, CT.

Sherman, P. (1970). In "Industrial Rheology," p. 370, Academic Press, London.

Stanley, D. W.; Pearsen, G .P.; Coxworth, V. E. (1971). *J. Food Sci. 36,* 256-260.

Szczesniak, A. S. (1963). *J. Food Sci. 28,* 385-389.

Szczesniak, A. S. (1973). "Texture Measurements of Foods" (A. Kramer and A. S. Szczesniak, eds.), pp. 71-108. D. Reidel Publ. Co., Dordrecht, Holland.

Szczesniak, A. S. (1975). *J. Text. Studies 6,* 139-156.

Szczesniak, A. S. (1977a). *Food Technol. 31,* 71-75, 90.

Szczesniak, A. S. (1977b). *J. Text. Studies 8,* 119-133.

Turro, E. J. and Sipos, E. (1968). *Baker's Dig. 42,* 44-50, 61.

Wolf, W. J. (1970). *J. Agr. Food Chem. 18,* 969-976.

Wolf, W. and Baker, F. L. (1972). *Cereal Sci. Today 17,* 124-126, 129-130.

Wolf, W. and Baker, F. L. (1975). *Cereal Chem. 52,* 387-396.

Wolf, W. J. (1976). "Advances in Cereal Science and Technology" (Y. Pomeranz, ed.), pp. 325-377. Am. Assoc. Cereal Chem. Inc., St. Paul, MN.

Wood, J. C. (1967). *Food Manuf. 42,* 11, 12-15.

Yatsu, L. Y. (1965). *J. Cell. Biol. 25,* 193-199.

Yatsu, L. Y. and Jacks, T. J. (1972). *Plant Physiol. 49,* 937-943.

DISCUSSION

O'Dell: Can you give us any idea about the relative proportion of the two types of proteins in the seed, the conditions used to extract them, and their solubility?

Martinez: These characteristics will differ with the oilseed. Which of the oilseeds are you interested in - soybeans?

O'Dell: Yes.

Martinez: I would like to call on Dr. Walter Wolf of the
Northern Regional Research Center to respond to your question
on soybeans.

Wolf: The protein bodies account for at least 70% of the
proteins in the soybean. These structures contain primarily
the 7S and 11S fractions and these proteins can be coextracted
with the nonstorage proteins with water from undenatured, defatted flakes. The 7S and 11S proteins are globulins and insoluble at pH 4.5. Hence, they are easily recovered by acid
precipitation as presently practiced in the manufacture of soy
protein isolates. The nonprotein body proteins are soluble at
pH 4.5 and remain in the whey.

Wilding: I think that bringing up nutritional value as a
functional property is a very good point, and it should be
discussed a little bit more.
    Would you comment further about the difference between the
high molecular weight and low molecular weight protein
components of soy as it relates to functionality.

Martinez: As you know, I am more familiar with cottonseed
than soy proteins. But the literature on soy proteins suggests that the situation is similar with respect to functionality. The high molecular weight proteins are the proteins
that provide the so-called "functional" characteristics.
These are the proteins that dissociate under diverse condi-
tions--acid, strong alkali, heat, and reassociate to produce
textures that are specific to the conditions under which they
are formed. These proteins are therefore the key to gel formation and heat setability. The latter are the type of phenomena that provide the desired "functionality" in, for example, meat emulsions.
    In the preparation of meat emulsions, mechanical work is
used to reduce particle size of the fat globules and to distribute the globules throughout a protein matrix. The heat
sensitivity of the storage proteins is then utilized to stabilize the emulsion. The success of such an emulsion, therefore, is not dependent upon the classical emulsifying capacity
of the protein, but rather upon its ability to gel at a specific temperature and stabilize the fat particles. Thus, it
is the "functionality" of these high molecular weight, storage
proteins that forms the basis for utilization of soy proteins
by the food industry.

In turn, the low molecular weight proteins are important because of their potential interference with functional properties. I believe, and this is only conjecture, that the characteristics of the low molecular weight proteins of soy and cottonseed are similar. These proteins tend to be very high in sulfur amino acids and thus provide ready sites for interaction with the subunits of the high molecular weight proteins. Such interactions could totally change the functional characteristics and in many instances, eliminate the desired properties.

Wilding: I understand that there is a varietal difference in this respect. This is an important point to highlight because of the deficiency of these sulfur amino acids in soy protein. Would you like to comment on this?

Martinez: In USDA, we are approaching this problem in two ways. We are using the traditional selection procedures of plant breeders and are also examining the problem in terms of the basic plant physiology. Researchers at our Plant, Soil and Nutrition Laboratory in Ithaca, New York are using tissue culture of soybean cells to isolate and identify mutants with a higher than normal level of methionine. Once stable, high methionine mutants are obtained, the problem then becomes one of regeneration of plants from the best mutants.

This group is also utilizing the techniques of molecular-biology to identify the gene that controls production of the 11S protein in soy. Ultimately, it may be possible through gene manipulation to increase the proportion of 11S protein which, among the storage proteins, is relatively rich in the sulfur amino acids.

Wilcke: I think that problem is going to require some very basic research and I think Hymowitz has surveyed the existing cultivars of soybeans and he doesn't really find enough variation in methionine to encourage him very much, so the procedure you described is probably one that will be more productive.

SOY PROTEIN ISOLATES IN INFANT FEEDING

Samuel J. Fomon and Ekhard E. Ziegler

During the early months of life infants commonly derive
all or nearly all of their protein intake from human milk or
an infant formula. In the United States in 1975 it was estimated that approximately 10% of formula-fed infants were being
fed formulas with protein from soy protein isolate (Fomon,
1975), and there is reason to believe that a similar percentage receive such formulas at present. For some infants, a
formula based on soy protein isolate may serve as the sole
source of protein for at least six months. Assessment of the
adequacy of such formulas is therefore of considerable importance.

In this presentation we shall review reports of nitrogen
balance studies and growth studies of infants fed formulas
based on methionine-fortified soy protein isolates. Previously unpublished data from our unit will be summarized. In addition, we shall summarize results of a recent study in which
infants were fed a formula based on soy protein isolate and
did or did not receive a supplement of L-methionine.

I. ADEQUACY OF SOY PROTEIN ISOLATES FORTIFIED WITH METHIONINE

A. *Nitrogen Balance Studies*

When diets differing in source of protein are fed to
comparable groups of subjects under standardized conditions,
the finding of greater retentions of nitrogen per unit of nitrogen intake suggests better protein quality. For example,
in studies of infants between 5 and 10 months of age, Kaye et
al (1961) demonstrated greater retention of nitrogen from diets with milk protein than from diets with protein from cottonseed flour or peanut flour.

Two major precautions are necessary in interpreting
such comparisons of nitrogen balance. First, if the diets
differ in components other than protein (which is almost inev-

itable in studies of infants), one must consider the possibility that a difference in nitrogen balance may have resulted, at least in part, from the non-protein components of the diet. Second, failure to demonstrate a difference in nitrogen balance should not be interpreted as an indication of nutritional equivalency of the proteins (or diets) because nitrogen balance may not be a sufficiently sensitive criterion of protein quality.

Only a few nitrogen balance studies with infants or children fed formulas with protein from soy protein isolate have been reported: 6 balance studies with 5 subjects ranging from 11 to 32 months of age were reported by Graham et al (1970), and 6 studies with 6 infants from 114 to 118 days of age receiving a formula (5224A) providing 6.5% of energy as soy protein isolate were reported by Fomon et al (1973). In these studies no difference in nitrogen balance was demonstrated whether the feeding supplied protein from soy protein isolate (*Edi-Pro A*) supplemented with methionine or from cow milk. In the study by Fomon et al, an additional comparison with human milk was possible (Figure 1) and no difference was detected in nitrogen balance between infants fed the soy protein isolate formula and those fed human milk.

During the past ten years we have accumulated a considerable body of additional data on nitrogen balance studies with normal male and female infants fed formulas with protein from soy protein isolate (*Edi-Pro A* in all but one formula which contained *Edi-Pro N*). In the earlier studies these formulas were fortified with DL-methionine (6 mg of methionine per gram of protein) and in more recent studies with L-methionine (4-8 mg of methionine per gram of protein). Balance studies were performed according to methods in routine use in our unit (Fomon, 1974) and each subject received the designated feeding for at least 10 days before the balance study was carried out. Although soy protein isolate provided all or most of the dietary protein in these studies, most subjects consumed small amounts of other foods. As pointed out previously (Fomon et al, 1973), these foods are generally low in protein content and rarely contributed more than 10% of protein intake.

In Tables 1 and 2 and Figures 2 and 3 results are summarized and compared with similar observations of infants fed milk-based formulas. Twenty-three studies were performed with 8 infants 8 through 120 days of age receiving diets providing 8% to 11% of energy from soy protein isolate. Mean absorption of nitrogen (84.1% of intake) was similar to that observed in 157 balance studies with 34 infants fed milk-based formulas (Table 1). Mean retention of nitrogen was 165 mg/kg/day when soy protein isolate was fed and 160 mg/kg/day when cow milk

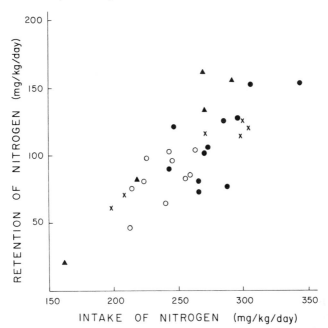

FIGURE 1. Retention of nitrogen in relation to intake of
nitrogen by normal infants between 90 and 120 days of age.
Each point refers to results of one 3-day metabolic balance
study. Each x refers to a study of an infant fed the soy-
isolate Formula 5224A. Other symbols refer to earlier observations: ▲ , infants fed a milk-based formula; o , infants
fed fresh human milk; ● , infants fed processed human milk
(reproduced from Fomon et al, 1973).

protein was fed. Because intake of nitrogen strongly influences retention of nitrogen, values for retention were corrected for intake by analysis of covariance and statistical
tests were applied to corrected values only. Mean corrected
retentions were 149 and 162 mg/kg/day, respectively, for the
two feeding groups; the difference was not statistically significant. The relation between intake and retention of nitrogen is indicated in Figure 2. It is apparent that the relation was similar for infants receiving soy protein isolate and
for those receiving cow milk protein.

    Twenty-nine balance studies were performed with 8 older
subjects receiving feedings providing 8-11% of energy from soy
protein isolate (Table 1). Absorption of nitrogen by these
subjects averaged 80.9% of intake; the corresponding value for
subjects fed milk-based formulas was 81.8%. Mean retentions
of nitrogen were 118 and 93 mg/kg/day, respectively, for the 2
feeding groups. Nitrogen retentions corrected for nitrogen
intake were 111 and 95 mg/kg/day, respectively; this differ-

Table 1. *Nitrogen balance studies with feedings providing 8-11% of calories from protein*

Protein Studies	Number of Studies	Number of Infants	Age[2] (days) mean (range)	Intake (mg/kg/day) mean (range)	Excretion Urine (mg/kg/day) mean	Excretion Feces mean	Absorption (% of intake) mean (S.D.)	Retention (mg/kg/day) mean (S.D.)	Retention (mg/kg/day) adj. mean[3]	Retention % of intake mean (S.D.)
Age of subjects: 8-120 days										
Soy[1]	23	8	70 (8-120)	456 (367-561)	218	72	84.1 (4.3)	165 (74)	149	35.6 (14.9)
Milk	157	34	58 (8-114)	430 (304-606)	203	64	85.1 (4.2)	160 (57)	162	36.6 (9.8)
Age of subjects: 121 days and older										
Soy[1]	29	8	283 (123-626)	368 (249-555)	180	69	80.9 (6.3)	118 (62)	111	30.9 (11.8)
Milk	100	24	264 (122-727)	350 (206-475)	194	62	81.8 (7.0)	93 (47)	95	25.6 (12.6)

[1] Soy protein isolate fortified with DL- or L-methionine.
[2] Age on first day of balance study.
[3] Corrected for intake by analysis of covariance using data for soy and milk.

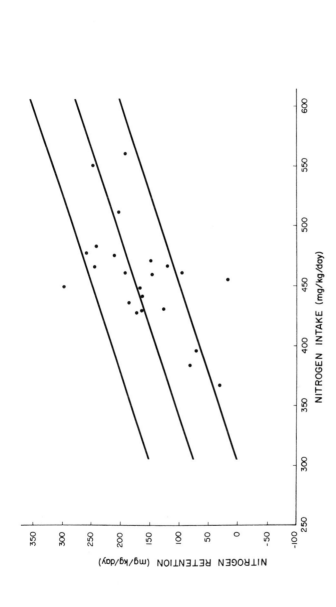

FIGURE 2. Retention of nitrogen in relation to intake of nitrogen by normal male and female infants 8 through 120 days of age receiving diets providing 8-11% of energy from protein.
Each dot refers to the result of one 3-day metabolic balance study with protein provided by
soy protein isolate. Lines indicate regression and 95% confidence limits for 157 balance
studies with formulas in which protein was provided by cow milk.

ence was statistically significant (t-test, p < 0.01). The
relation between intake and retention of nitrogen is indicated
in Figure 3.

Diets providing 12-16% of energy from soy protein isolate were fed during 36 balance studies with 8 infants 8
through 120 days of age (Table 2). Absorption of nitrogen averaged 85.8% of intake, a value similar to that observed with
infants receiving milk-based formulas. Mean retention of nitrogen was 163 mg/kg/day for infants fed protein from soy protein isolate and 226 mg/kg/day for infants fed milk protein.
Mean nitrogen retentions corrected for intake were 198 and 194
mg/kg/day, respectively; the difference was not statistically
significant.

Thirty-seven balances were performed with 12 subjects
125 through 764 days of age. Mean absorption of nitrogen by
subjects fed protein from soy protein isolate was 84.3% of intake. Mean retention was 160 mg/kg/day, equivalent to 32.7%
of intake. No suitable series of balance studies has been
performed with subjects in this age range receiving diets with
12% to 16% of energy supplied from cow milk protein.

In summary, available data on nitrogen balance studies
with infants fed soy protein isolates are limited to *Edi-Pro
A*. When protein supplied approximately 6% of energy intake,
nitrogen balance was similar whether protein was supplied from
methionine-fortified soy protein isolate, cow milk or human
milk. When protein supplied 8% to 11% of energy or 12% to 16%
of energy, nitrogen balance of infants 8 to 120 days of age
was similar whether they were fed formulas with protein from
methionine-fortified soy protein isolate or from cow milk. In
the case of older infants fed formulas in which protein supplied 8% to 11% of energy, retention of nitrogen was actually
greater when protein was supplied from methionine-fortified
soy protein isolate than when it was supplied from cow milk.
Thus, methionine-fortified soy protein (*Edi-Pro A*) appeared
at least as satisfactory as cow milk protein in promoting nitrogen retention.

B. Growth Studies

A number of reports have been published concerning
growth of infants fed formulas with protein from soy protein
isolate (Bates et al, 1968; Cherry et al, 1968; Cowan et al,
1969; Andrews and Cook, 1969; Graham et al, 1970; Wiseman,
1971; Sellars et al, 1971; Dean, 1973; Fomon et al, 1973; Jung
and Carr, 1977). These reports have been reviewed elsewhere
(Select Committee on GRAS Substances, in press). In several
of these reports the soy protein isolate is not identified and
in some instances important details about the subjects or
their management are not provided. Nevertheless, in all but

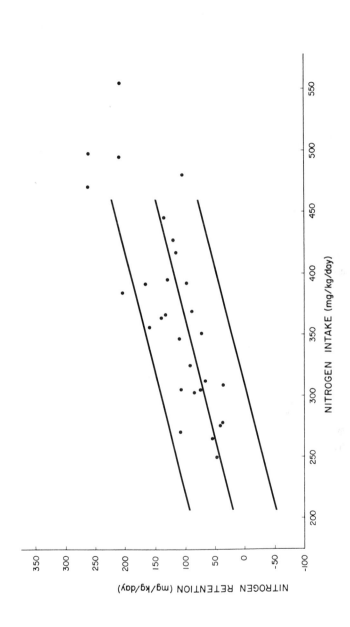

FIGURE 3. Retention of nitrogen in relation to intake of nitrogen by normal male and female subjects 121 through 727 days of age receiving diets providing 8-11% of energy from protein. Each dot refers to the result of one 3-day metabolic balance study with protein provided by soy protein isolate. Lines indicate regression and 95% confidence limits for 100 balance studies with formulas in which protein was provided by cow milk.

Table 2. Nitrogen balance studies with feedings providing 12-16% of calories from protein

Protein Studies	Number of Studies	Number of Infants	Age[2] (days) mean (range)	Intake (mg/kg/day) mean (range)	Excretion Urine (mg/kg/day) mean	Excretion Feces (mg/kg/day) mean	Absorption (% of intake) mean (S.D.)	Retention mg/kg/day mean (S.D.)	Retention adj. mean[3]	Retention % of intake mean (S.D.)
				Age of subjects: 8-120 days						
Soy[1]	36	8	62 (8-115)	559 (433-734)	317	79	85.8 (5.0)	163 (58)	198	28.7 (8.0)
Milk	38	8	56 (8-113)	658 (489-830)	355	82	87.8 (3.5)	226 (88)	194	33.8 (10.1)
				Age of subjects: 121 days and older						
Soy[1]	37	12	375 (125-764)	467 (275-628)	237	71	84.3 (5.7)	160 (84)	–	32.7 (14.5)

[1]Soy protein isolate fortified with DL- or L-methionine.
[2]Age on first day of balance study.
[3]Corrected for intake by analysis of covariance using data for soy and milk.

one (Cherry et al, 1968) of these studies, growth of the infants was interpreted as normal.

In the study by Cherry et al (1968), protein from *Promine K* supplied 15.1% of energy intake. The report fails to mention whether the formula was fortified with methionine. A control milk-based formula supplied nearly the same amount of protein (14.8% of energy intake). Seventy-three Negro infants were enrolled, usually by 3 days of age, and 58 completed the planned 6 months of study. Formulas were assigned alternately within narrow birth weight categories of like sex. Consumption of foods other than formulas was discouraged during the first 3 months of life and similar foods were consumed by both groups subsequently. By age 14 days mean body weights and lengths were greater for infants fed the milk-based formula than for those fed the formula with protein from soy protein isolate. In the case of boys, the difference in mean body weight was statistically significant at 42, 56, 80 and 96 days of age, but not thereafter, and the difference in length was statistically significant only at 96 days of age. In the case of girls, the difference in weight was statistically significant at ages 28 through 182 days and the difference in length was statistically significant at ages 28 through 152 days. Serum concentrations of total protein at most ages were slightly greater in the case of infants fed milk-based formulas, but the differences were not statistically significant.

In a study reported by Fomon et al (1973) 13 normal female infants were observed from 8 to 112 days of age while receiving a formula (5224A) providing 1.1 g of soy protein isolate (*Edi-Pro A*) per 100 ml (i.e., 6.5% of calories) fortified with L-methionine (5 mg/100 ml). Small amounts of other foods were permitted and these contributed an average of 7.7% and 6.1%, respectively, of the intake of energy and protein. Table 3 presents data on weight gain (per day and per 100 kcal) for various age intervals. Also presented are data from study of 10 female infants fed a milk-based formula with a comparable protein content (6.0% of calories) and of 116 breast-fed infants studied in the same fashion.

It is apparent that infants fed Formula 5224A demonstrated gains in weight similar to those of infants receiving the other feedings. In addition, serum concentrations of urea nitrogen, total protein and albumin were similar for infants fed Formula 5224A, Formula 29B and breast-fed infants. It was concluded that growth performance of infants fed Formula 5224A was similar to that of infants receiving comparably low protein intakes from other feedings.

Additional data, not previously published, from studies of a relatively large number of normal infants fed various formulas with protein from soy protein isolate (*Edi-Pro A*)

*Table 3. Mean gains in weight between 8 and 112 days of age
by normal female infants receiving various feedings*

	Number of Subjects	Age 8-42 days		Age 42-112 days		Age 8-112 days	
		(g/ day	(g/100 kcal)	(g/ day	(g/100 kcal)	(g/ day	(g/100 kcal)
Formula 5224A[2]	13	30 (6)[1]	7.1 (1.2)	23 (4)	4.3 (0.4)	25 (4)	5.1 (0.5)
Formula 29B[3]	10	30 (7)	7.1 (1.0)	22 (5)	4.2 (0.8)	24 (4)	5.1 (0.7)
Breast-fed[4]	116	34 (9)		22 (6)		26 (5)	

[1]*Numbers in parentheses indicate standard deviations.*
[2]*Data of Fomon et al (1973).*
[3]*Data of Fomon et al (1969).*
[4]*Data of Fomon et al (1978).*

fortified with DL- or L-methionine and providing 6.5 to 15
percent of calories from protein are summarized in Table 4.
These infants were observed from 8 to 112 days of age.
Seventy-four male infants fed 6 formulas and 67 female infants
fed 5 formulas were studied. Published (Fomon et al, 1971)
and unpublished data from similar studies of 174 male infants
and 159 female infants fed milk-based formulas are presented
for comparison. Intervals of study, characteristics of the
subjects and methods used for measurement of infants and for
determination of food intake have been described in detail
(Fomon et al, 1971). Male infants fed formulas with protein
from soy protein isolate demonstrated somewhat greater energy
intakes than did those fed milk-based formulas. In addition,
female infants fed formulas with protein from soy protein isolate demonstrated a slightly lower mean rate of weight gain
than did female infants fed milk-based formulas. These differences were not statistically significant. Mean weight gain
per 100 kilocalories, however, was significantly less
($p < 0.01$) during the observation period from 8 to 112 days of
age for both male and female infants fed formulas with protein
from soy protein isolate. Table 5 indicates weight gain per
unit energy consumed over shorter age intervals. The differ-

Table 4. Summary of results of growth studies with normal infants age 8 through 112 days[1]

	Males				Females			
	Milk		Soy Isolate		Milk		Soy Isolate	
Protein source	12		6		12		5	
Number of feedings								
Number of infants	174		74		159		67	
	Mean	S.D.	Mean	S.D.	Mean	S.D.	Mean	S.D.
Food consumption								
Volume of intake (ml/day)	826	92	861	106	746	81	760	82
Energy intake (kcal/day)	561	63	583	72	505	55	515	56
(kcal/kg/day)	106	9	109	10	104	10	108	10
Growth								
Weight gain (g/day)	32.2	5.6	32.3	5.9	27.5	4.8	26.8	4.5
(g/100 kcal)	5.74	0.71	5.52	0.68	5.44	0.74	5.19	0.59
Length gain (mm/day)	1.13	0.11	1.17	0.12	1.04	0.09	1.05	0.09

[1]Published (Fomon et al, 1971) and unpublished data.

Table 5. Weight gain per unit of energy consumed (g/100 kcal)

	Males				Females			
	(174)[1]		(74)		(159)		(67)	
Age	Protein Source				Protein Source			
Intervals	Milk		Soy Isolate		Milk		Soy Isolate	
	Mean	S.D.	Mean	S.D.	Mean	S.D.	Mean	S.D.
8-14 days	7.74	3.61	7.20	3.68	8.19[2]	3.60	7.03	3.81
14-28 days	8.89	1.67	8.53	1.91	8.32	1.66	7.96	1.40
28-42 days	7.38[2]	1.83	6.85	1.48	6.49	1.57	6.26	1.43
42-56 days	6.20	1.63	6.00	1.44	5.72	1.37	5.52	1.82
56-84 days	5.07	1.04	4.83	1.10	4.84[2]	0.96	4.50	1.05
84-112 days	3.95	1.05	3.94	0.98	3.91	1.27	3.85	0.84
8-42 days	8.03[3]	1.20	7.59	1.38	7.46[2]	1.28	7.06	1.18
42-112 days	4.82	0.81	4.69	0.71	4.61	0.76	4.44	0.63
8-112 days	5.74[3]	0.71	5.52	0.68	5.44[3]	0.74	5.19	0.59

[1]Number of infants.
[2]Value significantly (p < 0.05) greater than corresponding value for soy isolate.
[3]Value significantly (p < 0.01) greater than corresponding value for soy isolate.

ence in gain per unit of energy consumed was statistically significant for male infants during the age intervals 28 to 42 and 8 to 42 days and for female infants during the age intervals 8 to 14, 8 to 42 and 56 to 84 days. Concentrations of serum urea nitrogen, total protein and albumin did not demonstrate statistically significant differences between infants fed formulas with protein from soy protein isolate and those fed milk-based formulas.

Interpretation of these findings is difficult. Formulas with protein from soy protein isolate differ from milk-based formulas in a number of important respects besides the

source of protein. In the studies referred to in Tables 4 and
5, sucrose or sucrose and corn syrup solids provided the carbohydrate in formulas with protein from soy protein isolate;
lactose provided all or most of the carbohydrate in the milk-
based formulas. Various mineral salts are added in the manufacture of soy-based formulas whereas most of the minerals of
milk-based formulas are provided as components of the fat-free
milk employed in their formulation. Thus, the type of protein
is but one of a number of possible reasons for the lower
weight gain per unit energy consumed by infants fed formulas
with protein from soy protein isolate.

The evidence available suggests that formulas providing
protein from methionine-supplemented soy protein isolate
(*Edi-Pro A*) promote growth of infants to the same extent as
do milk-based formulas. The greater energy intake per unit of
gain in weight by infants fed soy-based than by those fed
milk-based formulas is of some interest and a matter for further study.

II. METHIONINE SUPPLEMENTATION

Although the quality of soy protein isolates as determined
by rat bioassay is unquestionably improved by the addition of
methionine, a similar effect in man has only recently been
demonstrated. Zezulka and Calloway (1976) studied 6 young men
receiving diets providing 9 g of nitrogen daily from one of 3
protein sources (egg white, *Promine A* and *Promine A* fortified with L-methionine) and made isonitrogenous (9 g of N/day)
by the addition of glycine and alanine. Nitrogen retention
increased with increasing intakes of nitrogen from soy protein
isolate (3.0, 4.5, 6.0 and 7.5 g/day) or from methionine-
fortified soy protein isolate (3.0, 4.5 and 6.0 g/ day). At
the same intake of intact protein, retention of nitrogen was
greater with methionine-fortified than with unfortified soy
protein isolate. When the intake of intact protein was 6.0 g/
day, retention of nitrogen was similar with the methionine-
fortified soy protein isolate and egg white protein.

We have recently completed a study designed to evaluate
the effect of methionine supplementation when infants receive
a diet providing moderate intakes of soy protein isolate
(Fomon et al, to be published). Growth performance and serum
chemical parameters were evaluated. Twenty normal male infants were fed a formula containing 1.56 g/100 ml (i.e., 9.3%
of calories) of a soy protein isolate (*Edi-Pro A*) and 19
completed the planned period of observation from 8 through 111
days of age. Of the infants completing the study, 10 received
a supplement of L-methionine in a dose of 5 mg per gram of
protein and 9 infants received no supplement. Methods of
study were as previously described (Fomon et al, 1971). The

Table 6. Food intake and growth of male infants fed Formula
          I-501 between 8 and 112 days of age

| | Methionine Supplementation | | | |
| | Without | | With | |
	Mean	S.D.	Mean	S.D.
Number of infants	9		10	
Food consumption				
Formula (ml/day)	774	127	816	71
Beikost (ml/day)	46	36	52	31
Energy intake				
(kcal/day)	564	74	597	60
(kcal/kg/day)	113	8	106	6
Protein intake				
(g/kg/day)	2.47	0.19	2.33	0.11
Gain in weight				
(g/day)	29.1	5.9	32.1	6.8
(g/100 kcal)	5.14	0.63	5.35	0.70
Gain in length				
(mm/day)	1.11	0.07	1.12	0.14

results of this study are summarized in Table 6. No statistically significant differences in any of the parameters of food
intake or growth were observed. Serum concentrations of albumin were similar in the two feeding groups (Figure 4). Serum
concentrations of urea nitrogen were significantly less in
supplemented than in non-supplemented infants (Figure 5). Because serum concentrations of urea nitrogen are inversely related to protein quality (Bodwell, 1977), this finding sug-

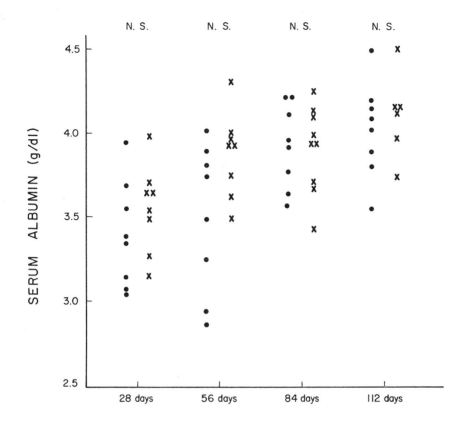

FIGURE 4. Serum concentrations of albumin at various ages of infants fed a soy isolate-based formula unsupplemented (•) or supplemented (x) with L-methionine. N.S. = not significant (t-test).

gests improvement in protein quality through supplementation.

Fourteen nitrogen balance studies were performed with 5 additional infants ranging in age from 67 to 307 days. Seven studies were performed with methionine supplementation and 7 without supplementation. As may be seen from Figure 6, methionine supplementation did not appear to influence retention of nitrogen.

SUMMARY

Published reports of studies of infants fed formulas with

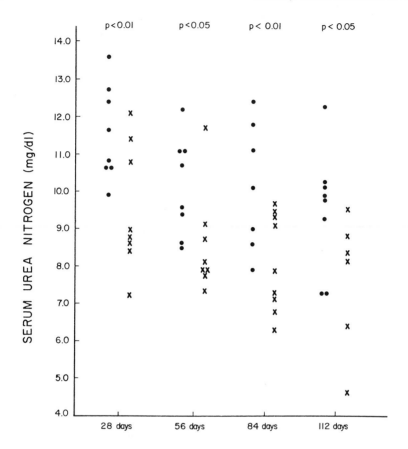

FIGURE 5. Serum concentrations of urea nitrogen at various ages of infants fed a soy isolate-based formula unsupplemented (●) or supplemented (x) with L-methionine. Significance levels (t-test) are included.

protein from soy protein isolate have been reviewed and additional data have been presented. Formulas with protein from
methionine-fortified soy protein isolate (Edi-Pro A) appear
to promote retention of nitrogen and growth to the same extent
as do milk-based formulas. It is of some interest that gain
in weight per unit of energy intake seems to be somewhat less
when soy-based formulas are fed than when milk-based formulas
are fed.

In studies of infants fed a formula providing 9.3% of energy from soy protein isolate (Edi-Pro A), the addition of
L-methionine did not appear to influence retention of nitro-

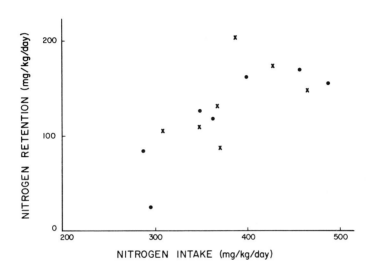

FIGURE 6. Retention of nitrogen in relation to intake of nitrogen by five normal infants receiving a soy isolate-based feeding (Formula I-501). Each point refers to the result of one 3-day metabolic balance study unsupplemented (●) or supplemented (x) with methionine.

gen, rate of growth or serum concentration of albumin. However, serum concentration of urea nitrogen was significantly less when the formula was supplemented with L-methionine, suggesting an improvement in protein quality through supplementation.

ACKNOWLEDGEMENT

The studies were supported by USPHS Grant HD 7578.

REFERENCES

Andrews, B. F.; and Cook, L. N. (1969). *Am. J. Clin. Nutr. 22,* 845–850.

Bates, R. D.; Barrett, W. W.; Anderson, D. W., Jr.; and Saperstein, S. (1968). *Ann. Allergy 26,* 577–583.

Bodwell, C. E. (1977). "Evaluation of Proteins for Humans," pp. 119–148. AVI Publishing Company, Westport, Connecticut.

Cherry, F. F.; Cooper, M. D.; Stewart, R. A.; and Platou, R. V. (1968). *Am. J. Dis. Child. 115,* 677–692.

Cowan, C. C., Jr.; Brownlee, R. C., Jr.; DeLoache, W. R.; Jackson, H. P.; and Matthews, J. P., Jr. (1969). *Southern Med. J. 62,* 389–393.

Dean, M. E. (1973). *Med. J. Aust. 1,* 1289–1293.

Fomon, S. J.; Filer, L. J., Jr.; Thomas, L. N.; Rogers, R. R.; and Proksch, A. M. (1969). *J. Nutr. 98,* 241–254.

Fomon, S. J.; Thomas, L. N.; Filer, L. J., Jr.; Ziegler, E. E.; and Leonard, M. T. (1971). *Acta Paediatr. Scand.* Suppl. 223.

Fomon, S. J.; Thomas, L. N.; Filer, L. J., Jr.; Anderson, T. A.; and Bergmann, K. E. (1973). *Acta Paediatr. Scand. 62,* 33–45.

Fomon, S. J. (1974). "Infant Nutrition," 2nd edition, p. 549. W. B. Saunders, Philadelphia.

Fomon, S. J. (1975). *Pediatrics 56,* 350–354.

Fomon, S. J.; Ziegler, E. E.; Filer, L. J., Jr.; Anderson, T. A.; Edwards, B. B.; and Nelson, S. E. (1978). *Acta Paediatr. Scand.* Suppl. 273.

Graham, G. G.; Placko, R. P.; Morales, E.; Acevedo, G.; and Cordano, A. (1970). *Am. J. Dis. Child. 120,* 419–423.

Jung, A. L.; and Carr, S. L. (1977). *Clin. Pediatr. 16,* 982–985.

Kaye, R.; Barness, L. A.; Valyasevi, A.; and Knapp, J. (1961). "Progress in Meeting Protein Needs of Infants and Preschool Children," pp. 297–312. National Academy of Sciences––National Research Council, Publication 843, Washington, D. C.

Select Committee on GRAS Substances (SCOGS 101) (in press). Life Sciences Research Office, Fed. Am. Soc. Exp. Biol., Bethesda, Maryland.

Sellars, W. A.; Halpern, S. R.; Johnson, R. B.; Anderson, D. W., Jr.; Saperstein, S.; and Shannon, B. S., Jr. (1971). *Ann. Allergy 29,* 126–134.

Wiseman, H. J. (1971). *Ann. Allergy 29,* 209–213.

Zezulka, A. Y., and Calloway, D. H. (1976). *J. Nutr. 106,* 212–221.

DISCUSSION

Bressani: The percentage of nitrogen retained was highest in infants fed human milk, followed by those fed cow's milk, and those fed the soybean product had the lowest percent retention.   Is it possible, from your data, to calculate the regression of nitrogen retention versus nitrogen intake and, by the regression analysis arrived at, a quality value such as the nitrogen balance indicated to compare the various protein sources?

Fomon: We did regressions - we have done I don't know how many hundreds of regression analyses; and, of course, when you do it you must only do it where you have feedings that cover the same range of intake.   Every time you compare soy formulas with milk-based formulas you are likely, if many commercially available formulas are included, to find that the intake of nitrogen will be greater with a soy-based formula, and that is not a fair comparison.   One cannot just pick out the values that overlap.   Formulas that provide the same amount of nitrogen must be used because otherwise there are other variables that enter in that give differences that are not related to the question under consideration.   At the 6 to 7-1/2% of the calories from protein, there is no difference that we can detect by any method of statistical analysis, and that's the level at which I think we have greatest sensitivity.   Now, we could have greater sensitivity if we fed lower intakes of protein, but we are not willing to do that.

Bressani: What about the digestibilities?   Have you looked at the protein digestibilities in these studies?

Fomon: Let me refer that question to my colleague, Ekhard Ziegler.

Ziegler: We have looked at absorption, that is net absorption that is determined during nitrogen balance, the same way we have looked at the retentions, and we cannot determine any difference between the milk-based and soy-based formulas.

Rotruck: In looking at your level of methionine supplementation, it didn't seem like very much to me.   If you calculate that based on total sulfur amino acids, it's about 12 to 13% of the total sulfur amino acids, I wonder if you considered whether or not a higher level of methionine might give an effect in your type of study.

Fomon: Let me comment first on why we chose that level of supplementation. We chose it because that's approximately the same level that is used in commercial infant formulas. It would be of no interest to use a higher level in comparing nitrogen balance because we already have with the lower levels, even without supplementation, the same retention to intake ratio that we have with milk-based protein. Now, it is possible that with a further supplementation with methionine, we might possibly show a difference on some of the other parameters. I would predict that the cystine in the plasma would be higher; I don't know if the urea nitrogen would be lower.

Zezulka: It seems that two of the different questions being addressed here are (1) at a good level of protein intake, such as 2-1/2 grams of protein at which you did the study with and without methionine, is there really a need for methionine, and (2) at lower levels will the methionine have an effect if the soy at low levels would really provide enough? If the methionine level requirements are really much lower than Snyderman's original data, which I think you have shown is at a lower level, the type of data that really needs to be looked at is at what level does soy then become limiting in methionine?

Fomon: Right. I agree that is the direction of our thinking as we design our further studies. We have a study going on now with a lower protein concentration being fed to somewhat older babies, and if we find nothing that suggests any difference in that study, we will proceed with that lower level in younger babies. Then, if the younger babies do not show it, we may even try an even lower level if we can get a satisfactory feeding. But, we are unwilling to proceed in large steps when our subjects are normal human beings.

Bodwell: This is in line with Dr. Zezulka's question, which is part of my question, too. If you go the other way, and go to a higher level of protein in the formulas as you say you normally would find in commercial formulas that are marketed, and look at L-methionine supplementation there, I wonder if you get a greater increase in the urea nitrogen in the unsupplemented, as compared to supplemented. Under very practical conditions, not looking at methionine requirements, what happens at this higher intake of soy protein as in the commercial formulas?

Fomon: I have no answer to that. My guess would be that the higher the protein intake, the less likely one would be to find an effect of sulfur amino acid supplementation.

Torun: In relation to one of the questions that was asked first, I would just like to say that we have been doing at INCAP a series of studies in amino acid requirements on 2- to 4-year-old children. We have found that the methionine requirements for children, at least of this age group, seem to be definitely lower than those currently accepted. Of course, the currently accepted values are interpolations between Snyderman's and Holt's data for infants and data for older children; but, for whatever it might be worth, I think it would be interesting for some people here that we have also found levels that seem to be lower in terms of requirements. Now, I would also like to make a comment and ask a question, Dr. Fomon. I find it very difficult to interpret the nitrogen balance data at levels of 400 to 500 mg nitrogen/kg/day of intake. Our experience has been with these levels of intake almost all proteins behave approximately in the same fashion as to nitrogen balance. Although I understand and realize that it is very difficult to go to very low levels with small infants, I think it would be very useful to look at some of the regression analysis data at lower levels, not as low as to jeopardize the children's well-being, but in the range where we can find differences within different proteins.

Fomon: I certainly accept the criticism. I would like to comment that to just remind you that the study by Kaye et al with the cottonseed protein and peanut protein compared with milk-based protein was within the same general range as the lower level of protein, the 8-11% of calories that we studied. That was where most of our data fell. I agree with you that the higher intake of protein, 12-14% or whatever it was, is too high. We would like very much to study more formulas in the 6 to 7-1/2% range of calories from protein, and we made at least three unsuccessful tries because the formulas were not satisfactory. It's just technologically very difficult and we just need more help from that point of view.

NUTRITIONAL QUALITY OF SOYBEAN PROTEIN ISOLATES:
STUDIES IN CHILDREN OF PRESCHOOL AGE

Benjamin Torun

The need for new sources of protein and food mixtures with
a high protein content for human nutrition has increased and
the trend towards their more widespread use seems to continue.
The higher availability and lower cost of vegetable proteins
compared with animal proteins make the former more suitable as
new sources or to increase the protein content or nutritional
value of existing foods. This can be achieved through mixtures of vegetables with complementary amino acid patterns
(Behar, 1963), through genetically improved vegetable proteins, such as opaque-2 corn (Mertz et al, 1964; Concon, 1966)
or through vegetable proteins with a good native amino acid
score, such as soybean protein. Digestibility must also be
taken into account when vegetable proteins are considered for
human consumption (FAO/WHO, 1975.).
One of the nutritional applications of these protein
sources is in the supplementary feeding of children. This and
the growing interest in soybeans for human nutrition in Central America prompted the testing of their protein quality in
children of preschool age. Towards this end two soybean protein isolates with low trypsin inhibiting activity were
used[1]. Table 1 shows their chemical composition and amino
acid pattern and Table 2 shows their content of essential amino acids and chemical score relative to the FAO/WHO reference
(1973). *Supro 620* has high viscosity and gel-forming properties and *Supro 710* has lower viscosity, non-gelling properties and is more easily dispersible in water.

[1]Ralston Purina Company, St. Louis, Missouri.

Table 1. Physical, chemical and nutritional characteristics
of soybean protein isolate[1]

Physical Properties   *Amino Acid Content (%)[2]*

Color	Cream	Alanine	3.46
Flavor	Bland	Arginine	6.39
Odor	None	Aspartic acid	9.32
pH (water slurry)	7.1	Cystine	1.37
		Glutamic acid	16.03
Chemical Composition		Glycine	3.46
		Histidine	2.21
Protein (N x 6.25)	85.2%	Isoleucine[3]	3.97
Fat (acid hydrolysis)	5.4%	Leucine[3]	6.63
Fat (ether extract)	1.1%	Lysine[3]	5.40
Fiber (crude)	0.1%	Methionine[3]	1.14
Ash	3.7%	Phenylalanine[3]	4.43
Moisture	5.3%	Proline	4.49
Calcium	0.2%	Serine	4.38
Phosphorus	0.8%	Threonine[3]	3.21
Sodium	1.3%	Tryptophan[3]	1.17
Potassium	0.05%	Tyrosine	3.11
		Valine[3]	3.98
Other Characteristics		Ammonia	1.73
Urease, pH increase	0.01		
Residual trypsin inhibitor, units/mg	7.5		
PER	2.10		
Casein corrected PER	1.63		

[1]*Data for* Supro 620. *Characteristics of* Supro 710
were similar.
   [2]*g/100 g of product.*
   [3]*Essential amino acids.*

EXPERIMENTAL DESIGN AND PROCEDURES

        Nitrogen (N) balance techniques were used to evaluate the
protein quality of the isolates (Allison, 1955). Using regression analysis, the regression coefficients between N intake or absorption and N retention of the test protein were
compared with those of milk as a reference protein. The regression coefficient of intake on retention provides a measure

Table 2. Essential amino acid content and score of the two soybean protein isolates

Amino Acid	Supro 620		Supro 710	
	Content[1]	Score[2]	Content[1]	Score[2]
Isoleucine	4.66	117	4.93	124
Leucine	7.78	111	8.15	117
Lysine	6.34	115	6.48	117
Methionine + Cystine	2.95	84	2.75	78
Phenylalanine + Tyrosine	8.85	148	9.41	157
Threonine	3.77	94	3.78	94
Tryptophan	1.37	137	1.38	138
Valine	4.67	93	4.87	97

[1]*g amino acid/100 g protein.*
[2]*Relative to FAO/WHO reference, 1973.*

similar to net protein utilization, NPU. The regression coefficient of absorption on retention is a measure similar to the biological value, BV. In either case, the better the quality of the protein the greater the regression coefficient (Bressani and Viteri, 1970; Viteri and Bressani, 1972).

Apparent digestibility was also calculated and compared with that of the reference protein.

Two separate studies were carried out. *Supro 620* was the test protein in the first study. It was tested with descending levels of intake from 320 to 120 mg N/kg/day, equivalent to from 2.0 to 0.75 g soybean protein/kg/day (N x 6.25). Fifty mg of choline chloride were added daily to the diet in addition to the vitamin and mineral supplements shown in Table 3 based on reports which suggest that the need for choline may increase when methionine is the limiting amino acid, as is the case with soy protein (Du Vigneaud et al, 1941; Treadwell, 1948; Mudd and Poole, 1975). That amount of choline chloride is equivalent to the content of free choline in about 500 ml of cow's milk (Macy et al, 1953).

Table 3. Vitamin and mineral supplements administered daily

Vitamin A	2500 I.U.
Vitamin B_1	1 mg
Vitamin B_2	0.5 mg
Niacinamide	5 mg
Vitamin B_6	0.5 mg
Pantothenic acid	5 mg
Folic acid	30 mcg
Vitamin B_{12}	2 mcg
Biotin	50 mcg
Vitamin C	25 mg
Vitamin D	500 I.U.
Vitamin E	1.5 mg
Iron (as ferrous sulfate)	60 mg
Iodine (as KI)	100 mcg
Manganese sulfate	0.9 mg
Zinc sulfate	1 mg

Supro 710 was the test protein in the second study. Based on the results of the first study, it was tested with descending levels of intake ranging from 200 to 80 mg N/kg/day, equivalent to from 1.25 to 0.5 g soybean protein/kg/day (N x 6.25). Tests were carried out with and without the addition of choline to evaluate the effect of the vitamin, usually with the same children.

A liquid diet was used in the two studies. It was prepared with the soybean protein isolate, sugars, corn starch, peanut oil and salts. Table 4 shows a typical dietary formula

Table 4. Composition of daily liquid diet prepared for a 10-kilogram child. During the day it was divided in five equivalent meals.

Component	Amount	Protein	Fat	Energy
	g	g	g	kcal
Supro 620	22.2	20	1.2	91
Sucrose	100.0	–	–	400
Dextrins-Maltose (Mead Johnson, U.S.)	42.3	–	–	165
Corn starch	15.0	–	–	60
Peanut oil	32.1	–	32.1	284
Mineral mix[1]	6.1	–	–	–
Water[2]	582.3	–	–	–
TOTAL	800.0	20	33.3	1,000

[1]*6.1 g of mineral mix provides, as mEq: K^+ 60; Na^+ 10; Ca^{++} 10; Mg^{++} 4; Cl^- 60; $HPO_4^=$ 10; $CO_3^=$ 10; $SO_4^=$ 4.*
[2]*Approximately 10% of the water was flavored with cinnamon by boiling during 30 minutes around 40 g of cinnamon sticks in 1,000 ml of water.*

for a 10 kg child with an intake of 320 mg N/kg/day. The energy content of the diet was 100 kcal/kg/day, since that was the energy level used in previous experiments with milk (the reference protein source). The experience at the Institute of Nutrition of Central America and Panama (INCAP) has been that this energy intake allows rates of growth and weight gains at least as fast as those expected for children of the same size as the ones who participated in the present studies. The only exceptions were children ER and GR who received 90 and 110 kcal/kg/day, respectively, based on several weeks' observations of their dietary intakes and growth patterns which deviated from the other children with 100 kcal/kg/day. Some children also ate 150-190 g of fresh apple which provided 3-5 mg N/kg/day. The liquid formula was prepared weighing the ingredients with an accuracy of ± 0.1 g, mixing them with approximately 80% of the water required and boiling for 15 minutes. After the diet cooled, plain or cinnamon-flavored water was added to take it to the prescribed weight. It was homogenized with a blender and the amount for each meal was weighed in tared cups for each of the corresponding child's meals.

The studies were carried out with children aged 19 to 44 months who had been admitted to INCAP's Clinical Center for treatment for protein-energy malnutrition. All had recovered nutritionally at least one month prior to beginning the studies based on clinical evaluation, weight expected for height, creatinine-height index and hematological and other biochemical indicators (Viteri and Alvarado, 1970; Torun and Viteri, 1976; Viteri and Torun, 1978); and in the interim they ate a diet which provided about 2 g milk protein and 100 kcal/kg/day. Informed parental consent to participate was obtained, and the experimental protocols were reviewed critically and approved by a Committee on Human Rights and Participation of Humans as Experimental Subjects. Eight children participated in the first study and eight others in the second. Of the latter, seven received the protein isolate with and without additional choline, and one of them (child SB) received the protein isolate with choline on two separate occasions. Table 5 shows their ages, weights and heights. When the same child participated in the study more than once, it was at intervals of at least 2 weeks during which he received a diet with 1.5 g soybean protein, 0.5 g milk protein and 100 kcal/kg/day.

The diet was fed initially at the highest level of N prescribed (320 and 200 mg N/kg/day in experiments 1 and 2, respectively), and every 9 days the amount of soybean protein isolate was reduced by 40 mg N/kg/day. The energy intake was kept constant by adding sugar and/or corn starch to compensate for the decrease in protein isolate. The full cups were weighed immediately before each meal; the diet was given to the children, the cups rinsed with 20-40 ml of water, which the children also drank, and the empty cups were weighed again. The difference in weight after accounting for the rinse water was recorded as the amount of diet ingested. Some children vomited occasionally, usually shortly after finishing a meal. The vomit's weight was calculated by picking it up with pre-weighed towels and by weighing the children's pre-weighed bedding and clothes. It was assumed to be mainly liquid diet, and the same weight of food was given to the child at a later time on the same day to compensate for the vomit. Complete urine and fecal collections were obtained during the

Table 5. Age, height and weight of children who participated in studies with soybean protein isolates

	Supro 620 + Choline	Supro 710 + Choline	Supro 710
Initial			
Age, months	33 ± 7[1]	32 ± 8	31 ± 7
Height, cm	83.8 ± 5.7	84.6 ± 4.9	83.6 ± 4.0
Weight, kg	11.62 ± 1.01	12.58 ± 1.69	12.28 ± 1.41
Weight-for-height, % of expected	97 ± 4	106 ± 11	105 ± 9
Change[2]			
Weight, g/day	10.6 ± 2.8	7.2 ± 6.1	13.9 ± 5.6
Height, mm/month	9.3 ± 2.1	7.5 ± 2.4	9.1 ± 3.3
Weight-for-height, % of expected during experimental period[2]	4 ± 3	1 ± 2	3 ± 2

[1]Mean ± standard deviation, n = 8.
[2]Duration of experimental periods: 54 days for **Supro** 620 and 36 days for **Supro** 710.

last 4 of each 9-day period. All losses were accounted for by continuous surveillance of the children and weighing of bedding, clothes and diapers previously weighed or tared. Carmine red and charcoal were used as markers for the periods of fecal collection. Homogeneous aliquots of urine, feces and diet were analyzed for their N content.

In addition to the N balance techniques, the children were weighed nude every morning before breakfast and their height was measured on 2 consecutive days at 2-week intervals. Total plasma proteins were determined refractometrically at the end of every 9-day period from fingertip blood.

RESULTS

The diets were readily accepted and all children were in
good health throughout the studies, except for upper respira-
tory infections, apparently of viral origin, that required
symptomatic treatment. Child RR had a generalized rash without fever for two days followed by epidermal desquamation
while on *Supro 710* without choline. Five children had periods of moderate diarrhea which disappeared spontaneously without dietary changes or medication.

Growth was adequate in terms of weight and height (Table
5) and there were no consistent changes in the pattern of
weight gain when the protein content of the diets decreased.

Digestibility: Table 6 shows the apparent digestibility
of the protein isolates tested. It decreased with the lowest
N intakes, probably due to the relatively larger contribution
of metabolic (obligatory) fecal N losses. There were no differences between the soybean protein isolates, between the use
or non-use of choline, nor between the soybean isolates and
milk fed at the same levels of nitrogen (Scrimshaw et al,
1958; Bressani et al, 1958, 1963, 1967, 1969, 1972; Viteri and
Bressani, 1972).

Nitrogen balance: Table 7 and Figure 1 show the amounts
of nitrogen retained at various levels of protein intake. Two
different types of relationship between intake and retention
were observed in the study with *Supro 620*: there were no
significant changes in retention at intakes greater than 200
mg N/kg/day (r=0.272, p >0.1), whereas retention decreased
with intakes below that level. The comparisons of the linear
regressions with intakes ≤ 200 mg N/kg/day by analysis of covariance showed that there was no difference between the regression coefficients when either of the protein isolates was
used nor when choline was added. There was, however, a shift
to the left in the regression line of the second study (*Supro
710*) resulting in an apparent need of less dietary N to attain equilibrium (60 and 58 mg N/kg/day, with and without additional choline, respectively) than in the first study (86
mg/kg/day). The former figures were even lower than the intake of milk protein needed by similar children using the same
experimental techniques (68-84 mg N/kg/day) (Viteri and
Bressani, 1972; Bressani et al, 1972).

The regression equations of the pooled data for each dietary treatment were similar to the mean of the individual regression lines, as shown in Table 8.

Nitrogen balance index: The nitrogen balance index is
derived from the relationship between apparent (i.e., without
accounting for obligatory losses) nitrogen absorption and re

Table 6. Apparent digestibility (% of intake) of soybean protein isolates at various levels of nitrogen intake

N Intake[1] mg/kg/day	Supro 620 + Choline	Supro 710 + Choline	Supro 710	Milk[2]
320	86 ± 5 (87 ± 3)[3]	-	-	84
280	86 ± 7 (88 ± 3)	-	-	83
240	87 ± 3	-	-	82
200	87 ± 3 (88 ± 3)	83 ± 7 (85 + 6)	86 ± 4	83
160	84 ± 4 (85 ± 2)	81 ± 7 (82 ± 7)	85 ± 6	78
120	82 ± 3[4]	75 ± 12 (79 ± 9)	75 ± 10[5] (80 ± 4)	-
80	-	70 ± 14[5] (74 ± 8)	65 ± 12[4] (68 ± 8)	69

[1]Nitrogen intakes ± 3%. Eight children with each treatment.
[2]Calculated from studies at INCAP: Scrimshaw et al, 1958; Bressani
et al, 1958, 1963, 1967, 1969, 1972.
[3]% absorbed, mean ± standard deviation. Figures in parentheses:
excluding balance of one or two children with diarrhea.
[4,5]Differs from other levels of N intake (Student's paired "t" test):
[4] $p < 0.01$, [5] $p < 0.05$.

Table 7. Apparent nitrogen retention (intake-urinary N-
   fecal N) of soybean protein isolates, mg/kg/day, at
              various levels of nitrogen intake

N Intake[1] mg/kg/day	Supro 620 + Choline	Supro 710 + Choline	Supro 710
320	88 $\pm$ 27[2]	–	–
280	81 $\pm$ 26	–	–
240	81 $\pm$ 17	–	–
200	74 $\pm$ 8	80 $\pm$ 14	81 $\pm$ 21
160	49 $\pm$ 14[3]	65 $\pm$ 14[3]	62 $\pm$ 10
120	22 $\pm$ 18	35 $\pm$ 10	32 $\pm$ 16
80	–	11 $\pm$ 11	14 $\pm$ 12

[1]Nitrogen intakes $\pm$ 3%. Eight children with each
treatment.
[2]Nitrogen retention, mean $\pm$ standard deviation.
[3]Treatments differ (Student's "t" test), p < 0.05.

tention (Allison and Anderson, 1945; Hoffman and McNeil,
1949). Table 8 and Figure 2 show the linear regressions of
apparent retention on apparent absorption. The regression coefficients at absorption ≤190 mg/kg/day correspond to the
nitrogen balance index. The results were similar to those described for the relationships between intake and retention:
two types of relationship were observed with N absorptions
above or below 190 mg/kg/day, there were no differences between the regression coefficients of both protein isolates,
the addition of choline had no effect and there was a shift to
the left in the regression line with *Supro 710* relative to
Supro 620. The regression equations of the pooled data for
each dietary treatment were also similar to the means of the
individual regressions.
    Plasma proteins: Table 9 shows the concentrations of
plasma proteins. The analyses within each dietary treatment
(using Student's paired "t" test) showed that the concentration of proteins decreased with N intakes of 120 or 80 mg/kg/
day. There were no differences between the two protein iso-

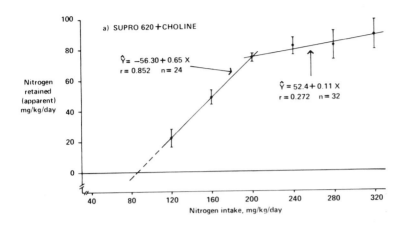

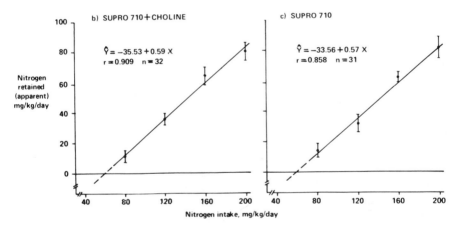

FIGURE 1: Relationship between nitrogen intake of soybean protein isolates and apparent nitrogen retention (intake-urinary N-fecal N), expressed as mg N/kg/day. The regression coefficients at intakes ≤ 200 mg N/kg/day do not differ. Points are means ± standard error (n=8, except 7 at intake of 80 mg N/kg/day in c).

lates nor with the addition of choline at any level of protein intake.

Table 8. Regression equations of apparent nitrogen retention (Y) on nitrogen intake (I) or absorption (A) of soybean protein isolates[1]

	Regression Equations of N Retention on	
	N Intake	N Absorption
Supro 620 + Choline	a^2	d^2
Pooled Data	Y = -56.3 + (0.65 x I)	Y = -47.2 + (0.70 x A)
Mean of Individual Regressions	Y = -56.3 + (0.65 x I)[3]	Y = -45.1 + (0.69 x A)
	(±41.8)[3] (+0.20)[3]	(+36.2) (+0.21)
Supro 710 + Choline	b^2	e^2
Pooled Data	Y = -35.5 + (0.59 x I)	Y = -25.2 + (0.66 x A)
Mean of Individual Regressions	Y = -35.5 + (0.60 x I)	Y = -24.1 + (0.65 x A)
	(±18.4) (+0.13)	(+9.6) (+0.08)
Supro 710	c^2	f^2
Pooled Data	Y = -33.6 + (0.57 x I)	Y = -19.8 + (0.59 x A)
Mean of Individual Regressions	Y = -33.6 + (0.58 x I)	Y = -19.6 + (0.59 x A)
	(±15.9) (+0.13)	(+9.4) (+0.12)

[1] Data from intakes ≤ 200 mg/kg/day and absorptions < 190 mg/kg/day.
[2] Analysis of covariance: no difference between regression coefficients in a, b, c nor in d, e, f. The residual variance of the coefficient in e was smaller than in d and f (p < 0.01).
[3] Standard deviation of the individual intercepts and regression coefficients, respectively.

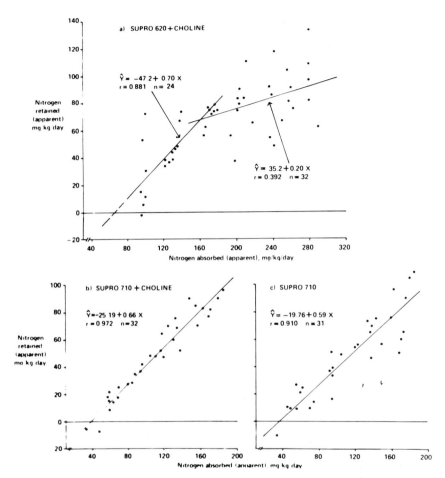

FIGURE 2. Relationship between apparent nitrogen absorption (without correcting for fecal obligatory N losses) of soybean protein isolates and apparent nitrogen retention (intake-urinary N-fecal N), expressed as mg N/kg/day. The regression coefficients at N absorptions below 190 mg/kg/day do not differ.

DISCUSSION

The results obtained with either of the two protein isolates were similar in terms of digestibility, nitrogen balance index and their effect on plasma protein concentration. The only differences observed were on the amount of N retained

*Table 9. Concentration of plasma proteins (g/100 ml) at
various intakes of soybean protein*

N Intake[1] mg/kg/day	Supro 620 + Choline	Supro 710 + Choline	Supro 710
320	6.9 + 0.2[2]	-	-
280	7.0 + 0.3	-	-
240	6.9 + 0.3	-	-
200	6.8 + 0.3	6.5 + 0.3	6.8 + 0.4
160	6.6 + 0.4	6.4 + 0.3	6.3 + 0.2
120	6.4 + 0.5[3]	6.3 + 0.5	6.3 + 0.3[4]
80	-	6.0 + 0.4[3]	5.9 + 0.3[3]

[1]*Nitrogen intakes + 3%. Eight children with each
treatment.*
[2]*Mean + standard deviation.*
[3]*Less than at other levels of intake (Student's paired
"t" test), p ≤ 0.05 or < 0.01.*
[4]*Less than with N intake of 200 mg/kg/day, p < 0.01.*

with intakes of 160 mg/kg/day (Table 7) and the intercept of
the regression equations which resulted in a displacement to
the left of the regression lines for *Supro 710* relative to
Supro 620. This, in turn, resulted in different estimates
of the dietary protein needed to attain nitrogen equilibrium
(Figures 1-3). To this end, intakes of *Supro 710* equivalent
to 58-60 mg N/kg/day would be needed in contrast to 86 mg N/
kg/day of *Supro 620* and 84 mg N/kg/day of milk protein, as
calculated by Viteri and Bressani (1972) from various studies
at INCAP using similar techniques with similar children.
There is no clear explanation for these apparently low requirements of *Supro 710*. A similar shift of the regression
intercepts of N retention on intake or absorption has been observed in dogs using the same protein sources on different occasions and it seemed to be related to the type of diet or nutritional status of the animals prior to beginning the balance
studies (Bressani, personal communication). The children who
participated in the present studies were equally well-

Table 10. Comparative results of nitrogen balance studies with soybean protein isolates and milk, at intake levels 200 mg N/kg/day

	Regression Coefficient Between N Retention[1] and		N Equilibrium Attained with N Intake[1] of	Compound (Mean) Score Relative to Milk
	N Intake[1]	N Absorbed[1]		
Whole Milk[2]	.64	.73	84	
Supro 620	.65 (101)[3]	.70 (96)	86 (98)	98
Supro 710	.59 (92)	.63 (86)	59 (142)	107

[1]*N intake, absorption and retention expressed as mg/kg/day.*
[2]*Source: Viteri and Bressani, 1972.*
[3]*Figures in parentheses: % relative to milk.*

nourished and ate similar diets before the experiments began. Those with *Supro 620*, however, had a slower and more gradual decrement in protein intakes since the experimental diets began with 320 mg N/kg/day, while *Supro 710* started with 200 mg N/kg/day. Most of the children studied with milk also received one or more dietary levels of intake higher than 200 mg N/kg/day. Another difference was that all children studied at INCAP with milk and with *Supro 620* were relatively sedentary while the children who ate *Supro 710* were stimulated to be physically more active. It has been shown that children who were recovering from protein-energy malnutrition with diets which provided 2.5 g protein and 120 kcal/kg/day grew better when they were physically more active, although there was no difference in N retention at that level of intake (Torun et al, 1975; Torun and Viteri, submitted for publication).

Based on those considerations it can be speculated that the children who participated in the second study had a decrease in obligatory fecal N losses or that they used "more efficiently" the protein fed at levels of 1.25 g/kg/day and lower, resulting in the displacement to the left of the regression lines of N retention on intake or absorption. The experimental evidence indicates that it is important to standardize the conditions used in studies based on N balance techniques, not only during the experiment itself, but prior

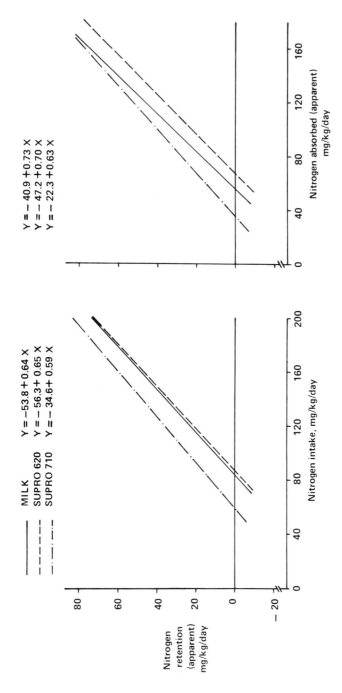

FIGURE 3. Relationship between nitrogen intake or apparent absorption (without correcting for fecal obligatory N losses) and apparent nitrogen retention (intake−urinary N−fecal N) of soybean protein isolates and whole milk.

to its beginning. It is possible that if depletion had taken
place at the same rate in the two studies and if there had
been no difference in physical activity, the results would
have been similar with both protein isolates. It is interesting to note that the effect was greater on the intercept than
on the regression coefficient. Therefore, the lack of unifor-
mity in the experimental conditions seems to affect the apparent requirement for a protein more than its biological value
relative to other proteins.

The addition of choline did not have any effects, at least
under the present experimental conditions and at the levels of
protein fed. In view of that and since seven of the eight
children studied ate *Supro 710* both with and without additional choline, the results of the two experimental treatments
with that protein isolate were combined. The new regression
equations of N retention on intake (I) and absorption (A),
with 63 data points, were Y= −34.6 + 0.59 I (r=0.884) and Y=
−22.3 + 0.63 A (r=0.939).

The two soy protein isolates studied were well tolerated
by the children and their digestibilities were comparable to
those of milk even when the balance periods with diarrhea were
included in the analysis (Table 6). Nitrogen retentions as
function of intake or absorption were compared with milk protein in Table 10 and Figure 3. Based on the nitrogen balance
index, the biological values of *Supro 620* and *Supro 710*
were 96% and 86% that of milk, respectively. Looking at the
regression coefficient of retention on intake as equivalent to
NPU, the soybean protein isolates had values of 101 and 92%
relative to milk. Finally, using the compound score suggested
by Viteri and Bressani (1972), which is the mean of the comparisons with a reference protein source of the regression coefficients of apparent N retention on intake and on apparent
absorption and of the N intake needed to attain N equilibrium,
the two soybean protein isolates scored 98 and 107% relative
to milk protein.

In conclusion, the soybean protein isolates tested compared quite adequately with milk based on nitrogen balance
techniques, ranging from 86% to 107% depending on the method
of comparison selected. Furthermore, all children retained
adequate amounts of nitrogen to allow for insensible losses
and growth requirements with intakes of 160 mg N (equivalent
to 1 g protein) per kilogram per day (FAO/WHO, 1973). This
level of intake coincided with that at which the concentrations of plasma proteins were maintained and, again, compares
satisfactorily with the recommended intakes of milk protein
for children of this age (FAO/WHO, 1973). Nevertheless, it
must be borne in mind that it is not practical to think of
soybeans as the sole protein source for preschool-age chil-

dren. But their high nutritional value makes them attractive
to complement other proteins of lower quality or to substitute
partially for others of equally high quality. However, the
nutritional value of such protein combinations must be assessed in each specific case, either theoretically (e.g.,
based on amino acid scores and digestibility (FAO/WHO, 1975)
or experimentally. The latter is, of course, more accurate
since theoretical estimates do not take into consideration individual variability nor the specific conditions of the experimental evaluation, such as those illustrated in the present
studies.

ACKNOWLEDGEMENTS

The author wishes to express his gratitude to the children
who participated in these studies and to the staff of INCAP's
Clinical Center. Mrs. Carolina Mena de Godinez played an important role in the studies as did Drs. Patricio Aycinena and
Salvador Garcia. Dr. Fernando Viteri contributed significantly in the experimental design and in the interpretation of the
results. The Ralston Purina Company kindly provided the materials needed for these investigations.

REFERENCES

Allison, J. B. (1955). *Physiol. Rev. 35:* 664-700.
Allison, J. B., and Anderson, J. A. (1945). *J. Nutr. 29:*
413-420.
Behar, M. (1963). *J. Amer. Med. Wom. Assn. 18:* 384-388.
Bressani, R.; Alvarado, J.; and Viteri, F. (1969). *Arch.
Latinoam. Nutr. 19:* 129-140.
Bressani, R.,; Scrimshaw, N. S.; Behar, M.; and Viteri, F.
(1958). *J. Nutr. 66:* 501-513.
Bressani, R., and Viteri, F. (1970). *Proc. 3rd. Int.
Congr. Food Sci. Technol., Washington, D. C.,* pp. 344-357.
Bressani, R.; Viteri, F.; Elias, L. G.; Zaghi, S.;
Alvarado, J.; and Odell, A. D. (1967). *J. Nutr. 93:* 349-360.
Bressani, R.; Viteri, F.; Wilson, D.; and Alvarado, J.
(1972). *Arch. Latinoam. Nutr. 22:* 227-241.
Bressani, R.; Wilson, D.; Chung, M.; Behar, M.; and
Scrimshaw, N. S. (1963). *J. Nutr. 80:* 80-84.
Concon, J. M. (1966). "Proceedings of the High Lysine
Corn Conference, Purdue University," pp. 67-73. Corn Indus-
tries Research Foundation, Washington, D. C.

Du Vigneaud, V.; Cohn, M.; Chandler, J. P.; Schenck, J. R.; and Simmonds, S. (1941). *J. Biol. Chem. 140*: 625-641.

FAO/WHO (1973). Energy and Protein Requirements. *Wld. Hlth. Org. Techn. Rep. Ser. No. 522.*

FAO/WHO (1975). *Guidance on the application of the 1973 Energy and Protein Requirements.* WHO NUT/75.4.

Hoffman, W. S., and McNeil, G. C. (1949). *J. Nutr. 38:* 311-343.

Macy, I. C.; Kelly, H. J.; and Sloan, R. E. (1953). "The Composition of Milks." *Nat. Acad. Sci. - Nat. Res. Council Publ. 254,* Washington, D. C.

Mudd, S. H., and Poole, J. R. (1975). *Metabolism 24:* 721-735.

Scrimshaw, N. S.; Bressani, R.; Behar, M.; and Viteri, F. (1958). *J. Nutr. 66:* 485-499.

Torun, B., and Viteri, F. (1976). *Rev. Col. Med. Guatem. 27:* 43-62.

Torun, B.; Schutz, Y.; Bradfield, R.; and Viteri, F. (1976). *Proc. 10th Int. Congr. Nutr. 1975,* pp. 247-249.

Treadwell, C. R. (1948). *J. Biol. Chem. 176:* 1141-1147.

Viteri, F., and Alvarado, J. (1970). *Pedriatrics 46:* 696-706.

Viteri, F. E., and Bressani, R. (1972). *Bull. Wld. Hlth. Org. 46:* 827-843.

Viteri, F., and Torun, B. (In press, 1978). "Modern Nutrition in Health and Disease," (R. S. Goodhart and M. E., Shils, eds.), 6th ed.

SOY PROTEIN IN ADULT HUMAN NUTRITION:
A REVIEW WITH NEW DATA

Nevin S. Scrimshaw and Vernon R. Young

INTRODUCTION

The soybean has long been an important food item of South-
east Asia in the form of bean curd and various indigenous fermented products. Relatively recently, however, it has been
considered a potentially important direct contributor to the
dietary protein intake of human populations that have not
utilized soy as a traditional food item. Moreover, a variety
of edible soy products, such as soy concentrates, isolates,
and texturized products fabricated with characteristics that
resemble various foods of animal origin have been developed
through advances in food technology. Further appraisal of the
potential of these products as nutritionally significant components of human diets is required.
      Proteins of plant origin tend to be of lower nutritive
value than those of animal origin because they are limiting in
one or more of the essential amino acids. This results in
poorer utilization and thus relatively more of the protein is
required if nutritional requirements are to be met. If, however, such deficiencies are corrected by adding the necessary
amino acids to improve quality, or by appropriate additions or
combinations with complementary protein sources, or through
genetic improvement of plant crops, vegetable proteins can
fully meet the dietary protein needs at levels of intake not
appreciably different from those required for animal protein
or a good mixed diet.
      Nutritional aspects of soy protein in human nutrition have
been reviewed by a number of investigators (Hamdy, 1974;
Bressani, 1975; Cater et al, 1978). This paper will be restricted to the recent literature concerned with the use of
soy protein concentrates or isolates to meet a significant
proportion of the daily protein requirements of human adults,

121

and to a summary of the results of unpublished MIT studies on
the nutritional value of soy protein isolates in adult humans.

AVAILABLE DATA ON SOY PROTEIN QUALITY FOR ADULT HUMANS

The primary objective of most biological assays concerned
with the nutritional evaluation of protein quality is to predict the value of a protein mixture or of individual proteins
as a source of nitrogen and essential amino acids for either
the animal species for which it is intended, or for humans.
However, the utilization of high-quality proteins is lower in
both children and adult humans when fed at requirement levels
than indicated by the standard assay procedures using young
rats (Young and Scrimshaw, 1978). For example, the utilization of egg or milk protein approaches 100% when measured at
the low intake levels commonly used for net protein utilization (NPU) measurements, but no more than 50-60% when measured
in human subjects at levels in the range of nitrogen equilibrium (Calloway and Margen, 1971; Young et al, 1973; Inoue et
al, 1973; Kishi et al, 1978.) This raises serious doubt as to
whether rat assays provide valid and quantitatively meaningful
information on the relative value of food proteins to support
long-term maintenance and growth of human subjects.

A revision of the NRC publication 1100, "Evaluation of
Protein Quality," prepared cooperatively by the International
Union of Nutritional Sciences (IUNS), the Protein Advisory
Group of the U.N. System (PAG), and the International Committee of the U.S. Food and Nutrition Board, now in press, concludes that neither PER nor the standardized NPU determination
is an adequate measure of the capacity of different food protein sources to support the nutritional needs of human beings.
Rat assay procedures using multiple levels of protein intake
are more reliable indicators of relative protein value. However, because of the apparently higher lysine and sulfur-
containing amino acid requirements of the young, rapidly growing rat, the appropriate biological measure of the capacity of
a food protein to meet human needs is obtained from studies
carried out in human subjects wherein the level of protein intake is at, or about at, the mean requirement level. A review
of the literature on the relation of animal to human assays of
protein quality emphasizes the inadequacy of existing information (Young and Scrimshaw, 1974, 1978), although the *relative*
value of various protein sources to meet protein requirements
of children may be determined with reasonable precision in the
growing rat (Bressani et al, 1973).

Clinical methods for the evaluation of protein quality are based on the same principles as those in the corresponding animal assays, but require some modification for application to human subjects. Growth in young children is sometimes used as a criterion, but the majority of studies that have been conducted in adult humans for determining the nutritional value of soy protein have been based on measurements of N balance. These may be supplemented with additional biochemical analyses, i.e., of serum proteins and amino acids, hemoglobin, blood urea nitrogen, and the urinary excretion of creatinine, sulfur compounds, and hydroxyproline (Young et al, 1977).

Kies and Fox (1971) reported the results of a study that utilized a textured vegetable protein (*TVP*) prepared by extrusion of defatted soy flour.[1] The objective of the study was to compare the protein value of *TVP*, methionine-enriched *TVP*, and ground beef when given at two levels of nitrogen intake to adult subjects. At the *lower* test level of dietary intake, 4.8 gm N/day (equivalent to about 0.4 gm protein/kg/day), mean nitrogen balances (gm N/day) were -0.30 for beef and -0.71 for *TVP*. Methionine addition improved N utilization of the *TVP* to an N balance of -0.45 gm N/day. Thus, the quality of beef appeared better than that of the unsupplemented *TVP*, but similar when the *TVP* was enriched with methionine. At a *higher and more adequate* test intake of total protein, 8.8 gm N/day (equivalent to about 0.7 gm protein/kg), the three sources resulted in almost identical nitrogen balances.

This same soy product was studied by Korslund et al (1973) with adolescent boys consuming the test protein at the inadequate 4.8 gm N level of protein intake. Although nitrogen retention was significantly better with the methionine-enriched *TVP* or beef compared with unsupplemented *TVP*, there was no significant difference in nitrogen balances of subjects fed beef or methionine-enriched *TVP*.

As sole sources of protein, soy products could be more or less limiting in methionine relative to human amino acid requirements, depending upon the composition of the product (Cheftel, 1977), and intake level of soy protein. This latter point is clearly demonstrated in the study by Zezulka and Calloway (1976a). These investigators assessed the minimum amount of soy protein with and without methionine required to meet amino acid needs in adult men consuming a diet supplying an adequate level of total nitrogen. Their study (Figure 1) showed that with methionine supplementation to bring the total

[1]Produced by Archer-Daniels-Midlands.

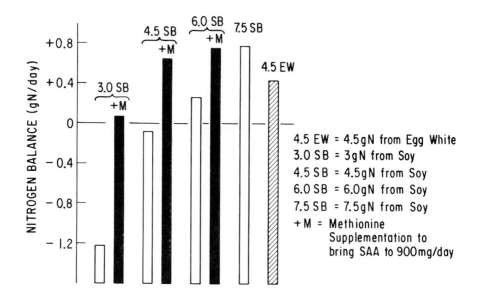

FIGURE 1. Nitrogen retention in young men receiving soy, with and without methionine supplementation, at various intakes, compared with egg white. Drawn from data of Zezulka and Calloway (1976a). Total diet supplied 9 gm N daily.

daily S-amino acid intake up to the level of 900 mg/day, stipulated by FAO/WHO (1973) to be sufficient for a 70 kg adult subject, the utilization of the soy protein isolate was improved at low intakes of soy protein.

At intakes of 6.5 gm soy nitrogen (equivalent to about 0.6 gm soy protein/kg/day), nitrogen balance was similar to that achieved with 4.5 gm egg white nitrogen (about 0.4 gm protein/kg/day), and the methionine requirement appears to have been met. Thus, the soy protein product was capable of meeting the amino acid requirements of adult subjects without supplementation when given as the sole source of dietary protein. However, supplementation with methionine improved the utilization of this product, making it possible to reduce the total amount of soy necessary to meet the essential amino acid requirements.

The use of crystalline methionine to supplement proteins limiting in S-amino acids presents a number of problems, including cost and the development of unacceptable flavors caused by methionine degradation. Hence, Zezulka and Calloway (1976b) have compared, as shown in Figure 2, the capacity of N-acetyl-L-methionine and of sodium sulfate to substitute for methionine as a supplement to soy protein. The former is nutritionally and metabolically equivalent to L-methionine (Boggs et al, 1975; Rotruck and Boggs, 1975), and at very high intakes, it may be less detrimental to rat growth than L-methionine itself is (Rotruck and Boggs, 1977). Even sodium sulfate produces a measurable and positive response.

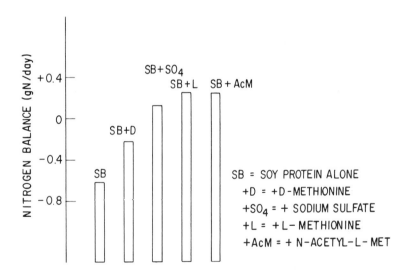

FIGURE 2. Mean nitrogen balance in young men receiving 4.5 gm N from soy protein isolate alone or supplemented with various sources of S-containing compounds. Drawn from data of Zezulka and Calloway (1976b).

MIT STUDIES OF SOY PROTEIN

The studies reviewed thus far provided useful comparative information on nitrogen utilization when one kind of soy protein product is consumed as the sole energy of dietary protein at limiting N intakes. However, assays involving more than one test protein level and, in particular, those that determine the N balance response to multiple doses of test protein intake within the submaintenance to maintenance range, are necessary for the critical quantitative determination of protein quality in human subjects, as well as in experimental animals (Scrimshaw and Young, 1974).

At MIT we have explored the nutritional quality of isolated soy protein in a series of such balance experiments in young adult males. The experimental methods and design of these studies, with egg or milk as reference protein sources, were comparable to those followed in our earlier work on the nutritional value of beef and wheat proteins (Young et al, 1975). We have now carried out eleven studies, to be reported in more detail elsewhere, with an isolated soy protein produced commercially by the Ralston Purina Co., St. Louis, Missouri, for food use.[2]

EVALUATION OF SOY PROTEIN QUALITY

Two different response criteria can be used to evaluate dietary protein quality from the N balance response curve, as shown in Figure 3 (Young et al, 1977). The first involves an estimation of the efficiency with which dietary nitrogen is utilized. This is determined from the slope of the nitrogen balance response curve in the region of the submaintenance-to-near-maintenance N intake level. Second, protein quality can also be estimated in relation to how well a given protein source meets the requirement for total protein and amino acids; i.e., what minimum intake level of a given protein is required to maintain a zero nitrogen balance or body N equilibrium. In this case, the response criterion is the intercept of the N balance response curve with the line of zero N balance.

[2]The specific products tested were isolated soy protein (*Supro 620*), widely used in meat application, and isolated soy protein (*Supro 710*).

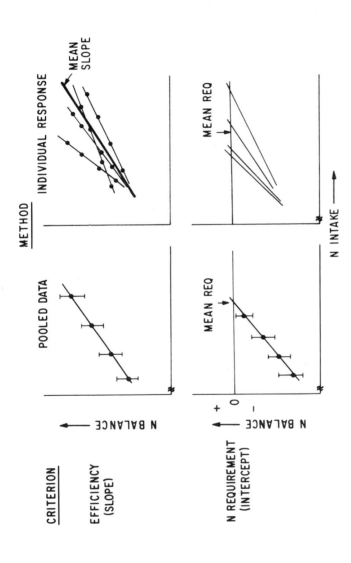

FIGURE 3. Schematic outline of two N balance response criteria and two methods of estimation for the assessment of dietary protein quality in adult human subjects receiving graded intakes of test protein within the submaintenance-to-near-maintenance level of intake. Modified from Young et al (1977).

These two approaches in the determination of dietary protein quality are interdependent but different. With them, the quality of dietary proteins may be compared using the relationship of the N balance responses with a test protein to those obtained with a reference or standard source such as egg or milk (FAO/WHO, 1973). Where the slope of test protein is compared with the standard, the assay is termed Relative Protein Value (RPV); where the intercept of the N balance response curve with the line of body N equilibrium is compared for test and standard protein, the assay is termed Relative Nitrogen Requirement (RNR). Two methods of estimating protein quality from these response criteria may be used.

The Pooled Data Method: For this method, the nitrogen balance response curve is measured at multiple levels of N intake in the submaintenance intake range. Standard least squares regression is used both to estimate the best straight line through the pooled N balance data from a number of subjects, and to calculate its variability. The slope of the line represents the efficiency of dietary N utilization, and in comparison with the reference protein, this provides the estimate of RPV. Alternatively, the intercept of this line with the line of body nitrogen equilibrium, or zero N balance, gives an estimate of the mean nitrogen requirement of the population and, in comparison with the reference protein, the RNR is determined.

The Individual Response Curve Method: This consists of analyzing the nitrogen balance data for each individual subject. Thus, a linear regression line is calculated to estimate each individual's nitrogen balance response. The mean efficiency of N utilization is the average value for the slope of the individual regression lines, and the individual nitrogen equilibrium values are used to calculate the mean nitrogen requirement of the group of individuals. When the sample is small, i.e., less than 10, this method exaggerates individual variation but is a more appropriate way of calculating the variation for larger samples; average intercept values are essentially the same in both methods.

The design of our initial study to compare the nutritional value of egg protein and isolated soy protein (*Supro 620*) is given in Table 1, and the resulting linear regression of N balance with intake is shown in Figure 4. Compared to egg, the Relative Nitrogen Requirement (RNR) was 80, based on analyses of pooled data, and the Relative Protein Value (RPV) derived from comparison of the slopes in this study was 87. On the basis of these mean values, the nutritional value of soy was only slightly less than that of egg protein. This was possibly related to the lower concentration of total S-containing amino acids in *Supro 620* compared with egg pro-

Table 1. Comparative Nutritive Value of Isolated Soy Protein 620 and Egg

Experimental Design

Subjects	15 Adult Males Aged 18 to 25 (8 on soy protein, 7 on egg protein)
Diets	Soy protein: 0.3, 0.4, 0.5, and 0.6 gm/kg/day Egg protein: 0.2, 0.3, 0.4, and 0.5 gm/kg/day Dietary energy: 38.4 to 55.1 kcal/kg/day
Periods	10-day dietary study: 1 day N-free diet 5-day adjustment period 5-day metabolic period randomized with 2-3 day *ad libitum* periods in between
Approach	Linear response of N balance with N intake with egg as the standard, comparison of slope (RPV) and N balance intercept (RNR)

tein. Therefore, we next carried out a series of studies to assess the effect of L-methionine supplementation on the utilization of soy protein in adult men.

Methionine supplementation in the first study was given to equal or somewhat exceed the level of the total S-amino acids suggested in the 1973 FAO/WHO Provisional Amino Acid Scoring Pattern (FAO/WHO, 1973). The Pattern provides 3.5% total S-amino acids. Methionine supplementation equivalent to 1.1% of test protein approximated this latter concentration, and the 1.6% level exceeded it. A 0.6% level of methionine supplementation was also tested. This study was conducted at a single level of test protein intake, approximately 0.51 gm soy protein/kg/day. This level was chosen to be close to the mean nitrogen requirement with good quality egg protein, but below the mean intake of soy protein found to be required for N equilibrium in our initial study. The results of this study are depicted in Figure 5, and they agree with those of Kies and Fox (1971) and Zezulka and Calloway (1976a). However, at the highest level (1.6%) of methionine supplementation, a statistically significant deterioration in overall nitrogen balance occurred.

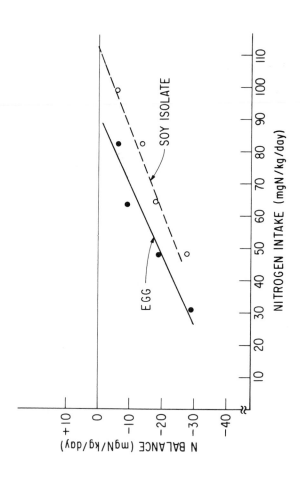

FIGURE 4. Mean nitrogen balances in young adult men receiving graded intakes of egg (seven men) or Soy Protein Isolate 620 (eight men). The lines for each protein were estimated by least squares analysis. Unpublished MIT studies.

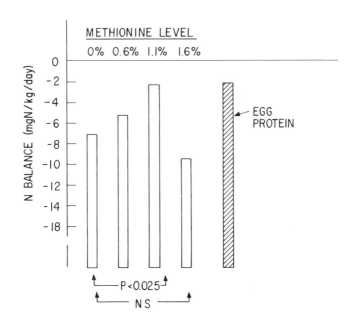

FIGURE 5. Mean nitrogen balance in young men given Soy Protein Isolate 620 supplemented with various levels of L-methionine. Supplementation expressed as a % of total protein intake (which was 0.51 gm protein/kg/day for all groups, including the egg protein control diet). The 1.1% level increased the total S-amino acid content to the level contained in the 1973 FAO/WHO Provisional Amino Acid Scoring Pattern (i.e., 35 mg total S-amino per gm protein). Unpublished MIT data.

This latter response parallels that reported in a series of INCAP studies in young children concerning the effects of amino acid supplementation of corn on nitrogen retention. In these studies (Bressani et al, 1958, 1960; Scrimshaw et al, 1958), not only did the addition of methionine fail to improve nitrogen balance when corn provided the entire source of di-

etary protein, but the addition of methionine to corn masa
protein at the level called for by the 1963 FAO/WHO reference
protein (FAO/WHO, 1965) resulted in a decrease in nitrogen
balance. Hence, N utilization in children is determined, in
part, by the dietary balance of essential amino acids, and a
disproportion in the amino acid pattern can adversely affect
dietary protein quality. Our study also indicates that at
marginal or deficient levels of total N intake, healthy young
men may also show an unfavorable metabolic response to relatively small changes or imbalances among some of the dietary
essential amino acids.

The finding of reduced quality with only a small excess of
methionine required confirmation and exploration of this phenomenon with a soy protein that intake was close to requirements for total nitrogen and essential amino acids. Therefore, two levels of methionine supplementation, equivalent to
1.1 and 1.6% of total soy protein intake, were studied for
protein intakes of 0.5 gm (about 5.6 g N/day) and 0.8 gm pro-
tein/kg/day (about 9.4 gm N/day). The results for both protein intakes are shown in Figure 6. For the lower protein
level, N balance improved with the 1.1% level of supplementation in six of eight subjects, and there was a reduction in N
balance with the 1.6% level of supplementation in these same
subjects. The other two subjects showed an improvement in N
utilization with the 1.6% level, but not at the 1.1% level.
Also shown in Figure 6, neither level of methionine supplementation resulted in an effect on N balance when isolated soy
protein intake was equivalent to 0.8 gm protein/kg/day.

Thus, methionine supplementation improves N balance in
young men when the level of soy protein intake supplies a limiting intake of S-amino acids, but an adverse response may occur with levels of methionine supplementation that do not
greatly exceed that estimated to meet the S-amino acid need
per unit of total N intake. However, when the level of soy
intake is sufficient to meet the dietary allowance for total
protein (e.g., Food and Nutrition Board, 1974), there is no
measurable effect of methionine supplementation.

From these studies, we conclude, in the context of the
usual diets of adults that include mixed sources of protein
such as meat, dairy products, and cereal grains, all of which
contain higher concentrations of S-amino acids relative to
soy, there is little nutritional or public health justification for requiring supplementation of soy protein products
with methionine.

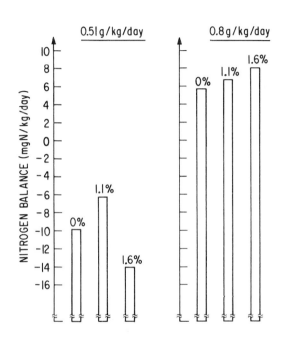

FIGURE 6. Mean nitrogen balances in young men given two
levels of Soy Protein Isolate 620 as the entire source of dietary nitrogen. Methionine supplementation was studied at
levels equivalent to 1.1 and 1.6% of total protein intake.
Unpublished MIT studies.

COMPARISON OF THE PROTEIN QUALITY OF SOY PRODUCTS AND BEEF

One important potential role of soy is as a meat extender
or substitute for animal protein sources in the diet of affluent societies.

In view of the Kies and co-workers findings discussed earlier (Kies and Fox, 1971; Korslund et al, 1973) that, *at limiting N intakes,* soy protein has a lower quality than beef in
adolescent boys and young adults, it is important to examine
further the extent to which soy may replace beef protein without an overall effect on the utilization of the dietary protein. To explore this point, Kies and Fox (1971) measured the

effect of different dietary ratios of beef protein to extruded
soy protein (*TVP*) on nitrogen utilization in eight young
adults. Their subjects received an intake of 4.8 gm N/day
(about 0.4 gm protein/kg/day) with a series of diets providing
a ratio of beef to soy of 4:0, 3:1, 1:1, 1:3, and 0:4. Their
results are depicted in Figure 7, showing nitrogen balances of
-0.44, -0.56, -0.75 -0.90, and -1.11 gm N/day, respectively,
for the foregoing dietary ratios of beef to soy nitrogen.
These data indicate a linear decrease in protein value of the
diet with increasing replacement of beef protein with the soy
product tested. However, in a subsequent study in which the
test level was 8.0 gm N/day (about 0.7 gm protein/kg), there
was no difference in N utilization of beef and soy. These results are also shown in Figure 7. Hence, it can be concluded

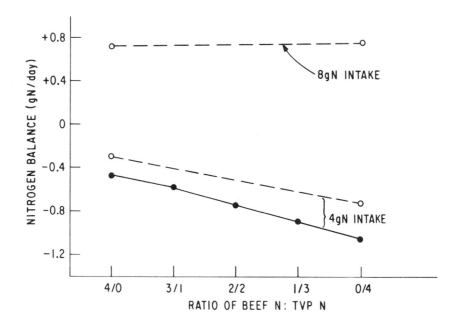

FIGURE 7. Nitrogen balance in young adults receiving
beef, or an extruded soy protein product (*TVP*) alone or in
various combinations. Taken from Kies and Fox (1973) (closed
circles), based in part on data of Kies and Fox (1971) (open
circles).

that the effects of various beef-soy combinations on N utilization will differ, depending on whether the measurements are
done at deficient or adequate levels of protein intake. Furthermore, the results would also be expected to depend on the
nutritional quality of the specific food product tested,
which, in turn, is influenced by the specific processing conditions used in producing the soy product.

For these reasons, we carried out a study to determine the
extent to which the high-quality *Supro 620* could replace beef
protein without significantly influencing overall dietary N
utilization. For this purpose, a single test level of 0.6 gm
protein/kg body weight/day was given. This was chosen on the
basis of a previous study indicating this to be the amount of
beef protein necessary in short-term balance studies to meet
the nitrogen needs of most young adult men (Young et al,
1975). It is also the level that Zezulka and Calloway (1976a)
found to be sufficient to achieve *mean* N balance in young
men receiving unsupplemented soy protein. Various combinations of beef and soy were tested, using a commercially prepared bologna product.

A summary of the N balance data obtained in this study is
presented in Table 2. Although significant (p 0.05) variability was found in nitrogen balances among the subjects,
there were no significant differences in nitrogen balance
among the various combinations of beef-to-soy protein. Thus,

*Table 2. Summary of N Balance Data and Protein Quality
           Indices in Soy-Beef Replacement Study*

% Soy	100	75	50	25	0
% Beef	0	25	50	75	100
N balance[1]	-2.3	-3.2	-0.9	-1.1	-1.7
BV[2]	53	52	55	53	53
Digestibility (%)	97	99	98	98	98

*None of the diets showed significant (P >0.05) differ-
ences.*
[1]*mg N/kg/day.*
[2]*Biological value.*

we conclude that a soy product of this quality, when given at a level equivalent to that sufficient to meet the requirement with beef protein, can replace beef without altering the protein value of the diet.

It should be emphasized that in our study, a 0.6 gm/kg/day level of test protein intake was given, equivalent to about 42 gm protein/day for a 70 kg man, compared with approximately 30 gm protein/day given to subjects by Kies and Fox (1971). This may be the major reason for the differences in results obtained in the two studies, but a difference in the nutritional quality of the soy products tested cannot be ruled out.

To explore further the protein value of combinations of soy and beef, we have compared beef, a 50% mixture (on an equal N basis) each of beef and isolated soy protein, and skim milk at test protein levels from 0.35 to 0.65 gm/kg. The study was carried out as before, with 10-day N balance periods, and seven subjects completed each diet group. The N balance results shown in Table 3 indicate no significant differences in the protein value of the soy-beef combination, compared with that for beef and milk.

The final MIT metabolic study to be summarized here is a direct comparison of the two different isolated soy protein products, *Supro 620* and *Supro 710*. The design was the same as employed previously, with eight subjects completing each of the trials. The data from the six subjects consuming

Table 3. N Balance Response in Young Men Receiving Beef, Soy-Beef (50:50), or Milk Protein

Source	Response Curve	Protein Intake for Balance (Gm Prot/kg)[1]
Beef	$\left[Y= -8.71+61.78 \ (x-.50)\right]$	$0.64 \pm .13$
Soy-Beef	$\left[Y= -4.56+46.94 \ (x-.49)\right]$	$0.59 \pm .12$
Milk	$\left[Y= -1.03+44.27 \ (x-.51)\right]$	$0.54 \pm .41$

[1]*0 balance assuming 5 mg N/kg integumental and miscellaneous losses.*
Differences among sources not significant.

milk were used both for direct comparison and pooled with previous similar data from other similar studies in which milk
protein was fed. Test protein intake levels were 0.35, 0.45,
0.55, and 0.65 gm protein/kg/day. The N balance data summarized in Table 4 indicate almost identical values among the
three sources of dietary protein.

Even at the lower protein intake levels, none has a consistent or significant advantage. When data from these last
two studies are combined, there are no significant differences
among these protein sources in the N required for estimated
zero balance.

From this series of studies, it is apparent that, in the
region of the minimum N requirement for N balance in young
men, the nutritive value of soy is the same as that for the
milk and beef proteins tested, and only slightly less than
that for egg protein. We had expected milk and egg to be of
identical value (e.g., FAO/WHO, 1973), but the consistency
with which milk protein appears to be required in somewhat
higher amounts compared with egg to achieve N equilibrium suggests a difference in the protein quality of these two
sources.

*Table 4. N Balance Response in Young Men Receiving Milk or
Soy Protein Isolates*

Source	Response Curve	Protein Intake for Balance (Gm Prot/kg)[1]
Soy protein 620	$\left[Y = -13.41 + 63.42 \ (x - .50)\right]$	$0.71 \pm .11$
Soy protein 710	$\left[Y = -11.57 + 66.68 \ (x - .50)\right]$	$0.68 \pm .35$
Milk	$\left[Y = -15.85 + 73.11 \ (x - .50)\right]$	$0.72 \pm .28$

[1]*0 balance assuming 5 mg N/kg integumental and miscellaneous losses.*
Differences among sources not significant.

LONG–TERM TOLERANCE AND ACCEPTABILITY OF SOY PROTEIN PRODUCTS

Even though soy protein is frequently used as a milk substitute for infants allergic to formulas based on cow's milk,
soy can also be a primary cause of allergy (Fries, 1966,
1971). Among atopic children, 5% had specific intracutaneous
reactivity to soy, while 64% had a positive reaction to two or
more members of the legume family, most often peanuts, string
beans, lima beans, and green beans. Processing is known to be
capable of producing allergenic substances in protein foods
not detectable in the original product (Scrimshaw 1975, 1978).
Thus, although allergic reactions are uncommon with traditionally processed soy-containing foods compared with many other
common foods of animal and plant origin, the possible effects
of new processing approaches should always be taken into account.

A comprehensive evaluation of the role of soy protein in
human nutrition must now include studies on the acceptability
of, and tolerance to, the long-term ingestion of the soy protein products that represent forms of preparation not previously experienced by man, or at levels of intake higher than
those on which human experience has been accumulated. These
studies must be conducted under close medical supervision and
supported by appropriate laboratory procedures.

In our research program thus far, we have conducted two
such acceptability studies. In the first study (Table 5), a
total of 40 healthy adult subjects were selected and random-
ized into two groups: One consisted of nine men and nine wom-
en consuming daily 40 gm of Soy Protein Isolate (*Supro 620*),
and a double-blind group of eleven men and eleven women who
consumed an equivalent amount of skim milk powder. The soy
and milk powder were incorporated into fruit juices, and were
consumed in addition to the free-choice diets of the subjects.
One subject in the experimental group, a 21-year-old woman,
developed mild gastrointestinal symptoms, nausea, and diarrhea
approximately two hours after consuming the material. Follow-
up tests suggested that the symptoms in this subject were associated with the ingestion of *Supro 620* and commercial
products containing it, but not with soy curd or *TVP* from
another manufacturer. The reaction was mild and, with repeated testing, the sensitivity gradually disappeared before a
specific cause could be identified.

The remaining subjects of both groups completed the study
uneventfully, without any adverse reaction of any kind that
could be attributed to the ingestion of the test product.
Blood samples were drawn initially and at 60 and 180 days of
the experimental period, and the results of a broad battery of

*Table 5. Design of a 6-Month Acceptability Study with an
Isolated Soy Protein (620)*

Design

Subjects: 40 adult subjects

(18 subjects - experimental group
22 subjects - control group)

Diets: experimental - 40 gm soy$_1$protein 620/day
control   - 40 gm DSM^1 powder

Length of
study: 6 months

Approach: clinical biochemical battery
physical examinations
reporting and appearance of clinical reactions

---

^{1}*Dried skim milk.*

*Table 6. Design of a 2-Month Acceptability Study with an
Isolated Soy Protein (710)*

Design

Subjects: 100 adult subjects
(50 experimental, 50 control)

Diets: experimental - 40 gm soy$_1$protein 710/day
control   - 40 gm DSM^1 powder

Length of
study: 2 months

Approach: clinical biochemical battery
physical examinations
reporting and appearance of clinical reactions

---

^{1}*Dried skim milk.*

blood biochemical determinations for the subjects in the experimental groups, relative to those in the control group, showed no changes or differences of clinical significance.

A second study was conducted on the acceptability of and tolerance to *Supro 710*. This study was carried out in a larger sample of subjects, as shown in Table 6. The subjects completed the study uneventfully and without significant complaints. Gastrointestinal functions remained normal throughout the study, and no nausea, abdominal discomfort, loose stools, constipation, or increased gas were reported. No allergic responses were observed, nor were there any changes in blood chemistries in the experimental group compared with those of the control group.

Kies and Fox (1973) monitored blood components of subjects consuming various textured soy protein products. All values for all subjects remained within the normal range during the experimental diet periods. Bressani et al (1967) showed, in their studies with children consuming another texturized food from soybean protein isolate, that the test product was readily accepted by all, and there were no adverse effects noted at any time during the experiment. Good acceptability of and tolerance to soy protein were also reported in studies concerned with protein quality of a corn-soy-milk mixture.

We have just completed a metabolic balance study in which *Supro 710* was the sole source of protein at a level of 0.8 gm/kg/day for eight male MIT students for *three* months of regular feeding. No clinical problems were encountered beyond the extreme monotony of the strict regimen, and there were no changes in blood biochemical values, including hemoglobin, serum proteins, and amino transferase enzyme activity.

Most reports of nutritional studies with soy protein products in humans provide no data on the acceptability of or tolerance to the material tested. If significant problems of intolerance had occurred, however, it is likely that they would have at least been mentioned. There seems no doubt that well processed soy products, when consumed at significant intake levels over long periods of time, are well tolerated and accepted by human subjects.

SUMMARY

This review has considered recent evaluations of processed soy protein in the nutrition of human adults. Nitrogen balance studies on the protein values of selected soy products for human use indicate that they are of good quality, approaching or equaling those of foods of animal origin, and are fully capable of meeting the essential amino acid and protein

needs of human adults. At total protein intakes that approximate current dietary allowances for good-quality protein in
adults, well processed soy protein isolate products fully replace beef without reducing the utilization of dietary nitrogen.

When the protein intake is inadequate, methionine supplementation improves soy protein utilization, but this is of
limited practical significance. There is considerable variability among subjects in their apparent requirements for S-
amino acids, and an adverse response to methionine supplementation was observed when the level of supplementation of soy,
given at a deficient protein intake, only slightly exceeded
the amount calculated to bring the total S-amino acid concentration up to that suggested by the 1973 FAO/WHO Provisional
Amino Acid Scoring Pattern. It is concluded that, under conditions of normal usage in adults, methionine supplementation
of good-quality soy protein products is unnecessary and probably undesirable.

The desirability of long-term studies concerned with tolerance to and acceptability of new soy protein products was
emphasized, and favorable results with two isolated soy protein products were reported. It is concluded that properly
processed soy protein foods are well tolerated and of good
protein value for human nutrition.

ACKNOWLEDGEMENTS

We thank Mr. Alan Wayler, Drs. Eduardo Queiroz and
Myriam Puig, and Ms. Edwina Murray for their key roles in the
conduct of the unpublished studies referred to in the text.
The financial support given by the Ralston Purina Company,
St. Louis, Missouri, and the help and advice of Drs.
C. Hardin, D. Hopkins, and F. Steinke are greatly appreciated.
Our studies utilized the facilities of the MIT Clinical Re-
search Center, supported by a grant (RR-88) from the Division
of Research Resources, National Institutes of Health.

REFERENCES

Boggs, R. W.; Rotruck, J. T.; and Damico, R. A. (1975).
J. Nutr. 105, 331-337.

Bressani, R. (1975). *J. Amer. Oil Chem. Soc. 52*, 254A-
262A.

Bressani, R.; Scrimshaw, N. S.; Behar, M.; and Viteri, F.
(1958). *J. Nutr. 66*, 501-513.

Bressani, R.; Viteri, F.; and Elias; L. G. (1973). "Proteins in Human Nutrition" (J. W. G. Porter and B. A. Rolls, eds.), pp. 293-316. Academic Press, New York.

Bressani, R.; Viteri, F.; Elias, L. G.; Zaghi, S.; Alvarado, J.; and O'Dell, A. D. (1967). *J. Nutr. 93*, 349-360.

Bressani, R.; Wilson, D. L.; Behar, M.; and Scrimshaw, N. S. (1960). *J. Nutr. 70*, 176-186.

Calloway, D. H., and Margen, S. (1971). *J. Nutr. 101*, 205-216.

Cater, C. M.; Cravens, W. W.; Horan, F. E.; Lewis, C. J.; Mattil, K. F.; and Williams, K. D. (1978). "Protein Resources and Technology: Status and Research Needs" (M. Milner, N. S. Scrimshaw, and D. I. C. Wang, eds.), pp. 278-301. AVI, Westport, Conn.

Cheftel, J. C. (1977). "Food Proteins" (J. R. Whitaker, and S. R. Tannenbaum, eds.), pp. 401-445. AVI, Westport, Conn.

Food and Nutrition Board, National Research Council (1974). "Recommended Dietary Allowances" (Eighth Revised Edition). National Academy of Sciences, Washington, D. C.

Fries, J. H. (1966). *J. Asthma Res. 3*, 209-211.

Fries, J. H. (1971). *Ann. Allerg. 29*, 1-7.

Hamdy, M. M. (1974). *J. Amer. Oil Chem. Soc. 51*, 85A-90A.

Inoue, G.; Fujita, Y.; and Niiyama, Y. (1973). *J. Nutr. 103*, 1673-1687.

Kishi, K.; Miyatani, S.; and Inoue, G. (1978). *J. Nutr. 108*, 658-669.

Kies, C., and Fox, H. M. (1971). *J. Food Sci. 36*, 841-845.

Kies, C., and Fox, H. M. (1973). *J. Food Sci. 38*, 1211-1213.

Korslund, M.; Kies, C.; and Fox, H. M. (1973). *J. Food Sci. 38*, 637-638.

Rotruck, J. T., and Boggs, R. W. (1975). *J. Nutr. 105*, 331-337.

Rotruck, J. T., and Boggs, R. W. (1977). *J. Nutr. 107*, 357-362.

Scrimshaw, N. S.; Bressani, R.; Behar, M.; and Viteri, F. (1958). *J. Nutr. 66*, 485-499.

Scrimshaw, N. S.; and Dillon, J. C. (1978). "Single-Cell Protein - Safety for Animal and Human Feeding." (S. Garattini, S. Paglialunga, and N. S. Scrimshaw, eds.), Pergamon Press, New York and Oxford, in press.

Scrimshaw, N. S. and Young, V. R. (1974). "Nutrients in Processed Foods - Protein" (P. L. White and D. C. Fletcher, eds.), pp. 109-120. Publishing Sciences Group, Inc., Acton, Mass.

Scrimshaw, N. S. (1975). "Single-Cell Protein II" (S. R. Tannenbaum and D. I. C. Wang, eds.), pp. 24-25. MIT Press, Cambridge, Mass.

World Health Organization (1965). "Protein Requirements" (Report of a Joint FAO/WHO Expert Group), Tech. Rep. Ser. 301, WHO, Geneva.

World Health Organization (1973). "Energy and Protein Requirements" (Report of a Joint FAO/WHO Ad Hoc Expert Committee), Tech. Rep. Ser. 522, WHO, Geneva.

Young, V. R.; Fajardo, L.; Murray, E.; Rand, W. M.; and Scrimshaw, N. S. (1975). *J. Nutr. 105*, 534-542.

Young, V. R.; Rand, W. M.; and Scrimshaw, N. S. (1977). *Cereal Chem. 54*, 929-948.

Young, V. R.; and Scrimshaw, N. S. (1974). "Nutrients in Processed Foods - Protein" (P. L. White and D. C. Fletcher, eds.), pp. 85-98. Publishing Sciences Group, Inc., Acton, Mass.

Young, V. R.; and Scrimshaw, N. S. (1978). "Protein Resources and Technology: Status and Research Needs" (M. Milner; N. S. Scrimshaw; and D. I. C. Wang, eds.), pp. 136-173. AVI, Westport, Conn.

Zezulka, A. Y., and Calloway, D. H. (1976a). *J. Nutr. 106*, 212-221.

Zezulka, A. Y., and Calloway, D. H. (1976b). *J. Nutr. 106*, 1286-1291.

DISCUSSION

Bodwell: Dr. Scrimshaw, in retrospect to Dr. Fomon's talk this morning have you measured plasma or urinary total urea nitrogen in your supplementation studies?

Scrimshaw: We have measured blood urea nitrogen (BUN) in all of our metabolic and tolerance studies, and have not observed clinically significant changes in these levels with soy protein feeding. However, there is a negative correlation between BUN and net protein utilization, as we have reported previously (Taylor, Scrimshaw, Young, Brit. J. Nutr. 32: 407 (1974).

Bodwell: The second question is on your acceptability or tolerance studies. Have you looked at blood lipid profiles and have you seen any differences?

Scrimshaw: Yes, in the long-term metabolic study just completed, 8 subjects consuming 0.8 gm per kg of *Supro 710* for 74-84 days showed significant decreases in serum cholesterol, triglycerides, and low density lipoproteins, with no change in high density or very low density lipoproteins. I omitted a comment from the summary, because it seemed beyond

the scope of this talk, to the effect that the increased use
of soy protein products in human diets either as substitutes
for meat, or to improve the protein value of predominantly
vegetable diets, could have positive health benefits beyond
the issue of protein value alone.

Bodwell: The third question is one that I am increasingly
concerned about when we use multiple protein intake levels for
evaluating nutritional quality. In adults at intake levels of
0.2 or 0.3 gram of protein per kilogram of body weight per
day, we observe marked negative nitrogen balances, minus 20 to
30 milligrams of nitrogen per kilogram body weight per day.
In assays with rats we always have positive nitrogen balance
if we have growth. In studies with children we try to keep
them in positive nitrogen balance. In our adult subjects at
levels of protein intake resulting in rather marked negative
nitrogen balances, are we measuring nutritional value? Or are
we measuring both differences in nutritional value and adaptation effects, thus compounding our estimates of nutritive value or protein requirements?

Scrimshaw: For evaluation of nutritional quality in young
adults, intakes of 0.2 to 0.3 gm protein per kg per day are
too low. This level results in a marked negative nitrogen
balance of between 20 to 30 mg N per kg per day. Rats are always in positive N balance if they grow. We try to keep children in positive N balance. So these rather distinct negative
nitrogen balances in adult subjects confound the estimates of
nutritional quality due to the marked adaptation to low nitrogen intake levels over short-term periods.
This is an important and valid point, particularly in relation to protein requirements and extrapolation of nitrogen
losses occurring on a nitrogen-free diet. Here, such individuals are measured at an extreme of adaptation in body nitrogen
metabolism, and represent a degree of conservation of body nitrogen quite impossible to sustain at requirement levels of
nitrogen intake. The basic criticism of NPU is that it involves measuring the utilization of protein at a level of adaptation that is simply not applicable at requirement levels.
Hence, one cannot predict precisely from the net protein utilization determined at the 0.3 gm/kg intake the intake of that
particular protein that would be required for maintenance of
nitrogen balance.
You will remember that the 1973 FAO/WHO Expert Committee
on Energy and Protein Requirements examined the evidence on
this point and concluded that it takes 30% more egg or milk
protein to achieve N balance when estimating the value at the

NPU standardized level as compared with that obtained at the requirement level. They assumed that the 30% figure applied to proteins of lower quality as well. We have published elsewhere that the poorer the quality of protein, particularly those limiting in lysine, the greater the discrepancy between the amount of a protein predicted from NPU standardized and the amount actually required to obtain nitrogen balance. With a protein as poor as wheat gluten, over 100% more is needed, too large an amount to consume (Young, V. R., *Proc. IXth Internat. Congr. Nutr.*, Vol. 3, pp. 348-362, Karger, Basel, 1975).

Bodwell: Yes, but my guess is that if we were able to obtain accurate values without such high individual variability, we would probably be seeing those curves taper off at the low protein intake levels just like Inoue has observed with wheat gluten. I'm just questioning if we aren't measuring adaptation effects as well as possible differences in nutritional quality.

Scrimshaw: First of all, the relationship of nitrogen retention with protein intake is not linear. It obviously is curved. We take an approximately linear portion of it. Second, 0.3 gm/kg protein intake is definitely too low, and this level was used in the earliest studies by Murlin, Hawley, and others. We do not use this intake as the single test level now because it cannot be linearly extrapolated to the zero balance line. Obviously, metabolic adaptations influence the position of the N-balance response curve as the intercept values are approached. If you go very much above requirement level in adults for points on the line, the intercept is shifted to excessively high intercept values. This is less of a problem in children or depleted adults; but still, if you go too much above, the slope of the line is distorted. Hence, it is essential to obtain data from several levels in the sub-maintenance to near-maintenance range without going too high or too low.

Hackler: When you are talking about a properly processed soy protein product are you talking about protein isolation procedures or processes, or are you talking about time/temperature processes, because they are quite different. When it comes to trypsin inhibitor content of soy products you can see a difference, depending on the processing or protein isolation procedures used.

Scrimshaw: I'm really not talking about processing at all. I am now speaking simply of the "black box" product that the food technologist gives to the nutritionist for testing. The nutritionist knows, of course, that the food technologist can, in one way or another, damage any protein by improper processing, and therefore must use caution in assuming that what is given to him has been well-processed. I meant no more than that.

Hackler: The second question has to do with the utilization of soy protein isolates by humans. If I understand your data correctly, when you have a protein product that is giving a PER of 1.5 to 1.8, such as your soy isolates, at least in terms of human adult nitrogen balance studies you have discussed, there is no difference in protein quality between that product (soy isolate) and say beef protein or milk protein. Therefore, would you suggest that we can lower the PER base of 2.50 for casein for purposes of nutritional labeling to a value of 1.5 to 1.8?

Scrimshaw: No. I suggest that PER should be discarded entirely as soon as possible. The AOAC should adopt a standardized slope ratio assay instead. One is proposed in the revision of the NRC Publication, "Evaluation of Protein Quality," now in press. It calls for an assay with about six animals per group at four levels of intake. Protein value is calculated from the slope of the regression line with the test protein relative to a standard such as casein or lactalbumin. This report presents overwhelming evidence that PER is not a suitable measure of protein quality for humans, and neither is NPU determined at a single deficient level of intake.

Hackler: I disagree; the slope is more expensive (it requires more labor, more animals and more dietary ingredients), and I would question that it is better for nutritional labeling purposes.

Wilding: I understood from your data that about .6 grams of protein per kilo per day is about the point where an adult comes into nitrogen balance. Is this correct, or was it more than that?

Scrimshaw: Well, I want to avoid giving any specific recommendation because a level that is sufficient in short-term balance studies may not be adequate for long-term maintenance. However, we have published studies in which egg protein was fed at 0.57 gm/kg body weight for 2-3 months with adequate energy intakes and have shown that the subjects lose lean body mass, as judged by creatinine excretion and K^{40} measurements, and an increase in serum transaminases (SGOT and SGPT), and a drop in serum albumin and hemoglobin (Garza, Scrimshaw, Young, J. Nutr. 107: 335, 1977). These changes are not necessarily reflected in nitrogen balance. Hence, we cannot assume from short-term balances in which .6 gm of isolated soy protein per kg appears adequate that it is an appropriate level for long-term balance.

Bressani: It's known from rat studies that the biological value of the same protein varies due to the kind of carbohydrate in their diet. Would not this also influence the value of the different proteins in food systems?

Scrimshaw: Certainly, we just completed similar metabolic balance studies comparing the different ratios of dietary fat to carbohydrate, and we are studying now the effects of different proportions of sucrose to complex sugars. We have found statistically significant differences in protein utilization in human adults under the former conditions. As Dr. Torun was saying this morning, there are many factors that can shift zero balance intercepts to the left or to the right. These include energy intake and balance, activity level, and the presence of infection, as well as the particular source of energy within the diet.

Bressani: Dr. Scrimshaw, I think that you have concluded that methionine supplementation was not needed for this type of protein foods, and I was wondering if the experimental approach used to measure this is the appropriate one. For example, when you were feeding .51 grams of protein per kilo per day and 1.67 of methionine there was a tremendous drop in nitrogen balance. This, of course, would be expected because of amino acid balance in the protein. These results reminded me of the work that you mentioned, on the addition of methionine to corn already supplemented with lysine and tryptophan. Now in this particular study, we were able to contrast the effect of methionine supplemention by the additon of threonine and valine. It is very peculiar that soybean protein has these amino acids, limiting amino acids on the second and third degrees after methionine, and wonder if a high level of addition did not induce deficiencies in these two amino acids. Could you comment maybe on the approach used to measure the significance of methionine supplementation of soybeans? Would it be better to use actually various levels of protein rather than to use only a high one or a low one as it was done for the beef and the blend of beef and soy protein?

Scrimshaw: You will notice that I chose my words very
carefully when I said "for healthy adults in the United
States." Where these materials might be used with individuals
whose dietary protein intakes may not be adequate, and who
need additonal protein for catch-up growth or recovery from
infection, the difference represented by a small addition of
methionine might have a significant effect. I was speaking
only in terms of the applicability of our kind of information
to food regulation in the United States. As Dr. Bressani
pointed out, and as many published studies illustrate, one can
certainly identify a variety of conditions in which methionine
supplementation improves nitrogen utilization. The most obvious of these is the use of a low protein intake of legume protein as the sole protein source. The question is: What conditions do you accept to be of practical significance for a
given situation?

We also need to have a further look into the issues raised
by Dr. Fomon this morning, particularly the differences in
BUN. We need to continue to look for other biochemical differences beyond those of nitrogen balance. I am not saying
for a moment that we should stop with, or be satisfied with,
nitrogen balance alone.

THE IMPORTANCE OF PROTEIN QUALITY
IN HUMAN NUTRITION

G. Richard Jansen

It is probably fair to say that the importance of protein
quality in human nutrition was considered better established
twenty years ago than it is today. Harper stated it well in
1962: "The efficiency with which the proteins of a diet are
used by a person depends upon how closely the quantities of
amino acids available from the food he consumes coincides with
the quantities required for his various body processes". According to this concept, proteins that closely meet the amino
acid needs of the body, such as egg, are of high quality and
are required in lesser amounts than poor quality proteins such
as cereals which are deficient in lysine. Utilizable protein
in a diet is, therefore, a function of protein quality and
quantity with improvements in protein nutriture possible
through increases in either or both of these parameters.
Since cereal proteins were generally considered to be of moderate to poor quality, it has been felt desirable to increase
the amount of utilizable protein of predominantly cereal diets. Approaches to this goal have included protein complementation, plant breeding for increased protein quality and amino
acid fortification.
The concept of protein nutrition described above has been
seriously questioned in recent years. Perhaps the most serious challenge to the importance of protein quality in human
nutrition has come from Payne (1975). In his words "the adequate safe level of protein-energy ratio in the diets of 2- to
3-year-old children is close to 5% and since most varieties of
cereal grains appear to provide utilizable protein levels of
close to this amount, this lends further support to the view
that primary protein deficiency is unlikely to be the main
factor causing protein-energy malnutrition in communities for
which cereals are the cheapest source of energy". Since protein quality is generally considered more important in children than in adults, and since cereal proteins are undoubtedly

149

Table 1. Protein Quality Studies in Monkeys and Kittens
          Studies in Young Monkeys[1]

Protein Source	Relative Potency
Growth	
Wheat gluten	15
Soy	46
Gluten plus lysine	48
Lactalbumin	100
Maintenance	
Wheat gluten	23
Soy	60
Gluten plus lysine	69
Lactalbumin	100

Studies in Growing Kittens[2]	
Protein Source	Weight gain (g/day)
Wheat gluten	1.8
Gluten plus lysine	11.6
Casein	15.9

[1]Data of Sammonds and Hegsted (1973)
[2]Data of Jansen, et al (1975)

the poorest quality of any food protein consumed in significant amounts by human populations, the view of Payne strongly
implies that protein quality is of no practical importance in
human nutrition except perhaps for infants in the first year
of life.

It is the purpose of this paper to assess the current situation as to the importance of protein quality. In this connection, I believe we can borrow from the German philosopher
Hegel. The importance given to protein in the fifties and
sixties can be considered the "thesis," and the challenge of

Payne and others in the seventies can be considered the "anti-thesis." We are currently in the early stages of what, again according to the analogy with Hegel, can be considered the "synthesis" phases.

To the extent that protein is important, it is for growth, nitrogen retention, support of pregnancy and lactation, recovery from illness and maintenance of health and well being. It is in these contexts that the literature will be reviewed. The emphasis will be on studies carried out in human subjects.

ANIMAL STUDIES

There is no doubt that the growth of rats or swine fed cereal or legume protein can be significantly increased by protein quality improvement, whether through plant breeding, amino acid fortification or protein complementation (Jansen, 1974; 1977; Bressani, 1977). Protein quality has also been demonstrated to be of considerable importance for growth of monkeys and kittens (Table 1). In young cebus monkeys, Samonds and Hegsted (1973) reported that the relative growth potencies for wheat gluten, soy protein, gluten plus lysine and lactalbumin were 15%, 46%, 48%, and 100% respectively. It is also significant that in this study, addition of lysine increased the relative potency of wheat gluten for tissue protein maintenance from 23% to 69%. Jansen, et al (1975) have shown that the weight gain of kittens fed a purified diet containing 29% wheat gluten can be increased from 1.8 to 11.6 grams/day by the addition of lysine monohydrochloride.

HUMAN STUDIES

Studies in humans in which protein quality has been improved are reasonably consistent with the animal work. Protein quality improvement of cereals or soy, whether by plant breeding, amino acid fortification, or protein complementation, increases growth and nitrogen retention in children and nitrogen retention in adults. A number of studies will be discussed here to illustrate the point, but the reader is referred elsewhere for more extensive reviews (Jansen, 1974; 1977; Bressani, 1977a; 1977b.)

Table 2. Fortification of White Flour with Lysine[1]
(% of Casein)

Case No.	None	Level of Added L-Lysine. HCL		
		0.28%	0.45%	0.90%
Weight Gain				
167	63	94	105	66
182	61	69	100	85
188	57	64	106	89
192	58	84	58	93
203	76	89	125	111
204	86	100	86	99
Mean	66.8+4.7[2]	83.3+5.8	96.7+9.3	90.5+6.1
Nitrogen Retention				
167	64	77	116	111
182	62	101	94	105
188	78	93	83	103
192	43	80	98	100
203	71	91	103	96
204	59	79	94	123
Mean	62.8+4.8	86.8+3.9	98.0+4.5	106.3+3.9

[1]Data of Graham, et al (1969)
[2]Mean + Standard Error

Studies in Infants and Children

Graham, et al (1969) evaluated the effects on growth and
nitrogen retention of the fortification of wheat flour fed to
infants ages 11-24 months at a protein level of 6.4-8.0% of
calories. The white flour was air classified and contained
21% protein. Levels of L-lysine monohydrochloride addition
were 0.28-0.90% equivalent to 0.1-0.4% lysine addition to
white flour with an 11% protein content. Weight gains and nitrogen balances were determined during three consecutive
three-day periods. As shown in Table 2, lysine addition significantly increased both weight gain and nitrogen retention
to levels equivalent to casein. In subsequent studies, these
authors (Graham, et al, 1971) demonstrated that lysine fortified white flour fed to infants for 3-6 months at protein lev-

Table 3. Methionine Supplementation of Soy Protein Studies in Infants[1]

Age (mo.)	Protein	DL-Methionine mg/Kg/day	Days	N Intake mg/Kg/day	N Retained mg/Kg/day
21	Sobee	--	6	240	42
	Sobee	11	12	232	70
23	Sobee	--	12	194	23
	Sobee	14	19	198	54
13	ProSobee[2]	--	9	320	95
	Sobee	--	15	320	79
13	ProSobee[2]	--	16	317	105
	Sobee	22	18	320	106
6	Soy Isolate	--	10	238	12
	Soy Isolate	20	9	240	44
8	Soy Isolate	--	6	240	63
	Soy Isolate	20	6	240	88
15	Soy Isolate	--	9	239	27
	Soy Isolate	20	9	242	57
21	Casein	--	11	320	107
	Soy Isolate	--	15	320	77
21	Casein	--	9	320	137
	Soy Isolate	16	6	320	107
19	Full Fat Soy Flour	--	32	280	69
	Full Fat Soy Flour	20	20	282	109

[1]Data of Graham (1971)
[2]ProSobee fortified with DL-Methionine

els ranging from 6.4-8.0% of calories supported normal linear
growth, weight gain, nitrogen retention, serum proteins and
liver morphology.

Reddy and Gupta (1974) compared opaque-2 corn and skim
milk in the treatment of twenty-two children ages 1-5 years
with kwashiorkor. Protein and calories were fed at 4 g/kg and
200 kcal/kg respectively. Parameters evaluated included the
time to lose edema, weight gain, plasma albumin increases and
blood hemoglobin changes. In this study, clinical and biochemical responses in the children fed opaque-2 corn were comparable to those in children fed skim milk. Although common
corn could not be used in this study for obvious ethical reasons, there is no doubt that common corn is unsatisfactory for
the rehabilitation of children with kwashiorkor. Bressani, et
al (1969) demonstrated almost ten years ago that the nitrogen
balance index of opaque-2 corn in children ages 2-5 is comparable to values obtained with milk.

The foregoing studies have been carried out with cereals
limiting in lysine. As shown in Table 3, addition of methionine to a variety of soy protein products increases nitrogen
retention in infants ages 6-23 months (Graham, 1971). In this
study, protein intakes ranged from 4.0 to 7.0% of calories.
As the percentage of calories in the form of protein is increased, a smaller response to methionine addition would be
expected. Fomon (1959) has reported that a soy formula without added methionine that supplied protein at 6.8% of calories
when fed to infants resulted in growth and nitrogen retention
equivalent to human milk.

Studies in Adults

The beneficial effects in human subjects on nitrogen retention of protein quality improvement of wheat by lysine addition were demonstrated by Hoffman and McNeil (1949) almost
30 years ago. More recently, Scrimshaw,. et al (1973) studied
the effects of nitrogen retention in sixteen young men of lysine additon to wheat gluten at adequate and restricted energy
intakes. As shown in Table 4, lysine consistently increased
nitrogen retention and decreased urinary urea at two levels of
protein intake, whether energy intake was adequate or not. In
later work, Young, et al (1975) estimated the relative protein
value of beef and wheat in young men to be 78+12 and 41+10
respectively, as compared to egg.

Several studies have been reported in which the effect on
nitrogen balance of methionine supplementation of soy protein
fed to adults has been evaluated. The results are summarized
in Table 5. Kies and Fox (1971) compared vegetable protein

Table 4. Lysine Fortification of Wheat
         Gluten in Young Adults[1]

Level of Protein (g/Kg/day)	Lysine	Urinary Urea (g/day)	N Balance (g/day)
Adequate Energy			
0.27	–	2.21 ± 0.44[2]	-1.29 ± 0.33
	+	2.04 ± 0.59	-1.05 ± 0.48
0.73	–	6.97 ± 0.80	-0.72 ± 0.73
	+	6.43 ± 0.92	-0.09 ± 0.69
Restricted Energy			
0.27	–	2.93 ± 0.58	-2.02 ± 0.49
	+	2.76 ± 0.99	-1.70 ± 0.96
0.73	–	7.53 ± 0.91	-1.40 ± 0.54
	+	7.31 ± 0.98	-0.97 ± 0.61

[1]*Data of Scrimshaw, et al (1973)*
[2]*Mean $\pm$ Standard Deviation*

with and without added methionine to beef at two levels of nitrogen intake. When fed at 4.8 grams nitrogen per kg, none of the protein sources resulted in positive nitrogen balance. Methionine fortification of textured soy protein reduced nitrogen loss. However, at a nitrogen intake of 8.8 grams/kg, methionine had no effect and nitrogen retentions on soy and beef were essentially identical.

In the studies of Zezulka and Calloway (1976), soybean protein isolate with and without methionine was compared with egg white. Methionine fortification increased nitrogen retention at nitrogen intakes ranging from 3.0 to 6.0 g/kg.

In summary, the studies that have been reviewed indicate that protein quality improvement of plant protein improves growth and/or nitrogen retention in young children as well as adults whether lysine or methionine is the limiting amino acid.

Table 5.   Methionine Fortification of Soy Protein in Adults

Protein Source	Methionine	N Intake (g/day)	N Balance (g/day)
Data of Kies and Fox (1971)			
Textured Vegetable Protein	–	4.8	-0.70
Textured Vegetable Protein	+	4.8	-0.45
Beef	–	4.8	-0.30
Textured Vegetable Protein	–	8.8	0.78
Textured Vegetable Protein	+	8.8	0.72
Beef	–	8.8	0.74
Data of Zezulka and Calloway (1976)			
Soybean Isolate	–	3.0	-1.21
Soybean Isolate	+	3.0	0.08
Egg White	–	3.0	-0.48
Soybean Isolate	–	4.5	-0.08
Soybean Isolate	+	4.5	0.64
Egg White	–	4.5	0.41
Soybean Isolate	–	6.0	0.26
Soybean Isolate	+	6.0	0.76
Egg White	–	6.0	0.73

Plasma Amino Acid Levels

Over the years, numerous attempts have been made to utilize fasting plasma free amino acid levels as an indicator of
protein quality with little success for lysine or methionine
deficient proteins (Young, et al, 1972; Graham, et al, 1972).
However, Longenecker and Hause (1959) pointed out nearly twenty years ago the potential utility of post-prandial changes in
free amino acids as an index of protein quality. These workers suggested that free amino acids are removed from the plasma in relation to need. Thus, the amino acid showing the
smallest increment or the greatest deficit when compared to
the requirement pattern would be the most limiting amino acid
in the dietary protein. Recent studies in human infants have
confirmed and expanded their original observations for proteins limiting in either lysine or methionine.

*Table 6. Post-Prandial Lysine as Indicator of[1]
          Protein Quality*

Parameter	Fasting	3 Hrs. Post-Prandial
Experiment 1 - Wheat		
Total Essentials (umoles/100ml)	74.1 ± 11.3[2]	74.8 ± 12.0
Lysine (umoles/100ml)	13.0 ± 2.0	10.2 ± 2.8
Threonine (umoles/100ml)	10.7 ± 2.9	11.1 ± 3.6
Experiment 2 - Wheat Protein/Casein (2:1)		
Total Essentials (umoles/100ml)	63.9 ± 10.7	78.6 ± 12.0
Lysine (umoles/100ml)	8.7 ± 2.2	10.7 ± 2.1
Threonine (umoles/100ml)	9.7 ± 4.5	11.5 ± 5.6

[1]*Data from MacLean, et al (1977). Protein fed at 6.4%
and 12.0% of energy respectively in Experiments 1 and 2.*
[2]*Mean $\pm$ Standard Deviation.*

MacLean, et al (1977) reported on the use of post-prandial
plasma lysine as an indicator of dietary lysine adequacy in
infants. A portion of their results are summarized in Table
6. When the dietary pattern was wheat, the post-prandial concentration of the total pool of free essential amino acids and
the concentration of threonine specifically showed no change
When compared to the fasting levels. In contrast, plasma free
lysine level was significantly lower in post-prandial plasma.
When the dietary protein was derived from wheat and casein
(2:1), plasma free lysine increased as much in the post-prandial plasma as did the total pool of free essential amino
acids.

The post-prandial amino acid levels also have been found

Table 7. Post-Prandial Methionine as Indicator of
              Protein Quality[1]

Parameter	Fasting	3 Hrs. Post-Prandial
Soy Protein		
Total Amino Acids (umoles/100ml)	255.0+35.4[2]	293.0+36.8
Essentials/Total	0.272+.025	0.267+.036
Lysine/Essentials	0.179+.029	0.175+.036
Cystine/Essentials	0.025+.010	0.028+.015
Methionine/Essentials	0.026+.005	0.022+.005
Soybean Protein with Methionine		
Total Amino Acids (umoles/100ml)	225.3+18.7	270.6+19.4
Essentials/Total	0.243+.017	0.247+.032
Lysine/Essentials	0.161+.032	0.172+.027
Cystine/Essentials	0.040+.018	0.036+.021
Methionine/Essentials	0.035+.006	0.051+.008

[1]Data of Graham et al (1976). Protein was fed at 6.4%
of energy.
[2]Mean + Standard Deviation

to be useful in documenting a dietary inadequacy in methionine. Graham, et al (1976) compared the plasma amino acids of
infants fed soybean protein with or without supplemental
methionine. A portion of their results has been summarized in
Table 7. When soy protein was fed, the level of total free
amino acids in plasma increased with the ratio of methionine
to the total amino acid pool showing a significant decrease,
as was the case for lysine with wheat. When methionine was
added to the soy, the ratio of methionine to the total amino
acid pool showed a substantial increase.

These results, although the post-prandial changes in plasma free amino acids were small, were obtained with levels of
dietary protein in the low to moderate range. The full significance of these changes in plasma free amino acids during
the post-prandial period is not fully understood at the pres-

Table 8. Effect of Protein Intake on Milk Secretion[1]

| | Protein Intake (g/day) | |
	50	100
Milk Solids (g/100ml)	13.8+1.3[2]	13.4+0.9
Milk Protein (g/100ml)	1.61+0.15	1.57+0.19
Milk Lactose (g/100ml)	8.1+0.9	7.9+1.0
Milk Produced (g/day)	742+16	872+32
Milk Consumed (g/day)	617+15	719+10
Infant Weight Gain (g/day)	30.4+3.6	45.7+2.0

[1]*Data from Edozien et al (1976). Energy intake was 60 Kcal/kg for each treatment (n=7)*
[2]*Mean + Standard Deviation*

ent time. However, it has been reported that the pattern of free amino acids in plasma is needed, not only to stimulate appetite, but also to provide the proper hormonal milieu and substrate supply to enable growth to occur (Leung, et al, 1969).

PROTEIN QUALITY DURING LACTATION

A time of heightened nutritional needs is during lactation. It is somewhat surprising, in view of the considerable stress placed in recent years on the importance of breast feeding, that relatively little work has been carried out on human nutritional requirements during lactation. Even less attention has been given to studying basic aspects of milk biosynthesis, especially as related to amino acid supply.

Lonnerdal et al (1976) compared breast milk composition and volume in poorly nourished Ethiopian mothers with the composition and volume of breast milk in well nourished Ethiopian and Swedish mothers. These workers found, as had several earlier studies, that neither the quality nor the quantity of breast milk was affected by maternal malnutrition. However, they emphasized that further studies are needed.

In a study carried out in Nigeria, Edozien and co-workers

Table 9. Protein Quality Improvement of Bread Fed at
              Different Energy Levels[1]

	White Bread (Ad Lib)	Bread + Lysine + Threonine (Ad Lib)	Bread + Lysine + Threonine (Restricted)
Food Consumption Pregnancy (g/day)	16.8+1.3[2]	19.9+0.5	14.6+0.8
Food Consumption Lactation (g/day)	31.4+1.4	38.5+0.9	27.2+0.1
Weaning Weight (g)	16.8+0.4	38.8+0.4	28.8+0.3
Brain Weight (g)	1.235+.010	1.491+.008	1.344+.012
Cerebellum Weight (g)	0.148+.003	0.177+.001	0.159+.001
Cerebrum Weight (g)	0.847+.012	0.966+.010	0.882+.008
Cerebellum Protein (mg)	11.90+.20	16.34+.10	14.26+.08
Cerebrum Protein (mg)	71.16+.99	81.12+.86	77.14+.70
Cerebellum DNA (mg)	1.023+.019	1.467+.010	1.144+.007
Cerebrum DNA (mg)	0.922+.016	1.088+.014	0.973+.013

[1]Data from Jansen and Monte (1977)
[2]Mean + Standard Error

(1976) evaluated the effect of a skim milk supplement containing 50 grams of protein on milk secretion. The energy
content of the basal diet was adjusted for the energy in the
supplement so that energy intake was maintained at 60 kcal/
kg. As shown in Table 8, the protein supplement caused a
significant increase in milk secretion and a 50% increase in
daily weight gain of the infants. It is clear that more work
is needed on the nutritional requirements for satisfactory
lactation in humans.

    In order to examine the effect of protein quality on lac-
tation, it is necessary to refer to the animal literature.

Boomgaart et al (1972) studied the effect of dietary lysine levels on lactation performance of first-litter sows. A corn-soybean diet furnishing 12% protein was fed during pregnancy. During lactation, the basal corn-soy-sesame meal diet furnished 16% protein and 0.6% lysine. Additional lysine did not improve weight gain of the offspring, leading the authors to conclude that the basal diet was adequate in protein and lysine. However, the maternal weight loss did appear to be reduced by higher levels of lysine. The authors speculated that second litter sows, because of their greater volume of milk secretion, would require a higher intake of amino acids than observed in this study. In any case, the lysine requirement for lactation in swine, as determined in this study, was 3.8% of protein. This is below the lysine level in the FAO scoring pattern, but is above the lysine content of most cereals.

In a recent study, Jansen and Monte (1977) reported on the effects of protein quality improvement of white bread fed at varying energy levels during gestation and lactation. Pregnant rats were fed, from conception until weaning, amino acid fortified or unfortified white bread at 100%, 85% or 70% of *ad libitum* consumption. A comparison of the most significant treatment groups is presented in Table 9. Lysine and threonine fortification was associated with an increase in weaning weight from 16.8+0.4 to 38.8+0.4 g, the latter weight being not significantly less than that observed on casein. Brain weight and brain DNA and protein in all regions of the brain were substantially increased. More significantly, even when the amount of the amino acid fortified bread diet fed was restricted so that mothers consumed 13% less protein and energy than did the mothers fed the unfortified bread diet, all parameters were significantly greater in the case where the higher quality protein was fed.

Because no similar studies have been carried out in a human population, we need to make the best judgment possible at this time concerning the implications for humans. A mother who would be secreting 800 ml of breast milk a day would secrete 640 mg of lysine a day in high quality milk protein. This represents a doubling of her lysine requirement over the nonpregnant, non-lactating situation. Under these conditions, it is difficult to see how lactating mothers consuming a poor quality cereal diet, even when supplemented with small amounts of legumes, would not benefit from protein quality improvement of the diet.

ASSESSMENT OF CURRENT SITUATION

Before considering whether protein quality improvement of
practical human diets would be of any real value, it is desirable to assess the current situation. Since there is little
evidence to suggest that protein quality is currently a
signficant nutritional factor in the United States and other
developed countries, the following comments are made in relation to nutritional problems in developing countries.
As is generally realized, protein-energy malnutrition of
young children is considered a very serious public health
problem in most third and fourth world countries. Typically
in less developed countries, there is a high rate of infant
and early childhood mortality and morbidity with the growth of
the survivors stunted. Birth weights are low and lactation,
when attempted, is often poor. To recognize the importance of
the socio-economic milieu and to realize that many non-nutri-
tional factors are involved is scant evidence that food quality is not a significant determinant.
The dietary situation is most developing countries, especially those where cereals represent the primary food staple,
can be characterized as follows:
Food energy consumed is marginal. Protein, as a percentage of calories, is reasonable but protein quality is marginal
as is the nutrient density of the diet in general. A very important factor, often ignored, is that most diets in, less developed countries are quite low in fat.

DIET IMPROVEMENT

It is fairly obvious that there are many social, economic
and agricultural factors that need to be addressed in developing countries, in order for long-term nutritional improvements
to occur. However, since the diet is quite often marginal in
amount and poor in quality, it would appear reasonable to conclude that more and better quality food has merit. There has
been much attention given over the last five years to the importance of food energy deficiencies in developing countries.
Protein-energy relationships have been considered in more detail elsewhere (Jansen, 1977; Scrimshaw, 1977) and will not be
further considered here.
Even if one grants that preschool children growing poorly
in developing countries do not consume enough food energy, one

is still faced with the problem of understanding why this is
so. It is clearly not always that food availability is limited by economic or cultural factors. Gershoff et al (1977) recently reported on the effects of a rice fortification program
in Thailand. The amino acid-vitamin-mineral fortified rice
and unfortified rice were fed to preschool children over a
several year period. Fortification did not significantly improve growth or morbidity of children. Two additional points
are significant, however, and could easily be lost sight of in
evaluating the implications of this study. The first is that
the children in both treatment groups grew poorly, and the
second is that this poor growth occurred in spite of the fact
that adequate quantities of traditional foods were available
to the children. The authors concluded that the children did
not meet their energy needs either because of the low calorie
density, poor palatability of their diets or some other unknown reason.

In view of the generally low level of fat in most diets in
developing countries, it would appear that more attention
should be given to increasing the fat content of such diets.
This would improve both calorie density and palatability.
However, as the calorie density of a diet is increased, the
protein energy level of the diet will decrease. Since protein
quality is already marginal, this suggests that protein quality will become even more limiting under these conditions. Animal foods have the advantage of increasing calorie density,
protein quality, protein quantity and palatability as well as
micronutrient intake and availability. However, economic factors will likely restrict the extent of diet improvement
through increases in animal food consumption. To the extent
that fat intake from plant sources, especially those rich in
essential fatty acids, can be increased, this would appear to
be highly desirable. Such an approach, using whole soybeans
or soy oil, for example, makes protein quality improvement of
predominantly cereal diets of increasing practical importance
as part of nutritional intervention programs in less developed
countries.

The data presented in this brief review demonstrate that
in an experimental situation human subjects, like other mammalian species, respond to dietary protein quality improvement
by increasing growth and nitrogen retention. The practical
questions are:
1) what, is the appropriate standard on which to judge protein quality and
2) are there any situations existing in the world where protein quality would appear to be lower than desirable.

These questions have been discussed more extensively elsewhere (Jansen, 1974, 1977). It would appear that humans, even

for growth, do not require as high a percentage of essential
amino acids in general, lysine and methionine in particular,
as are present in egg protein. A strong indication that this
is the case is that growth of young children on soy protein or
lysine fortified white flour is quite good, even when fed at a
relatively low percentage of protein calories (Graham, et al
1971; Fomon, 1959). On the other hand, cereal proteins in
general would appear to be too low in lysine to provide for
satisfactory growth and lactation. At the present time, the
most reasonable suggestion would appear to follow the FAO/WHO
scoring pattern (FAO/WHO, 1973) until a better guide comes
along.

Whether any practical human diets are lower in protein
quality than desirable would appear to depend primarily on the
percentage of calories supplied by cereals on the one hand and
legumes and animal foods on the other. Jansen et al (1977)
have recently examined the effect of geography and income on
amino acid patterns in Brazil. Amino acid scores based on the
FAO/WHO pattern were calculated for urban and rural populations in three geographic regions of the countries as a function of income. As shown in Table 10, in essentially all
cases, the limiting amino acids methionine and lysine were
present in the dietary patterns at greater than 90% of the
levels in the FAO/WHO pattern. The reason for this appeared
to be the increased consumption of beans among low income populations and animal foods in the upper income groups. In contrast, El Lozy et al (1975) calculated amino acid scores for
quantitative surveys of food consumption that had been carried
out in Southern Tunisia. In this situation, lysine was present at only 56% of that in the FAO/WHO pattern, not even considering lysine losses in cooking. Such a diet should benefit
by being improved in protein quality, whatever means are employed to accomplish this. It is correct to point out that a
field study of lysine fortification in Tunisia did not show
the expected benefit (Wilcke, 1977). However, in considering
the implications of this observation in relation to the importance of protein quality in human nutrition, two points are
often overlooked. The first is that no benefits were found
for vitamin and mineral fortification and yet this is not used
as a reason to not fortify cereals with these micronutrients.
The other point is that few nutritional interventions, by
themselves, have been demonstrated in the field to have had
the expected nutritional benefits. Rather than use the results of the fortification studies as justification for the
conclusion that man is a relatively unusual mammal where protein quality is unimportant, a better response might be to use
the knowledge gained to design a better experiment where the
concept of protein quality improvement of human diets might
again be put to the test.

Table 10. Amino Acid Scores of Dietary Patterns in Various Regions of Brazil[1]

Region	Income Level	Amino Acid Score							
		Lysine		Sulfur Amino Acid		Threonine		Tryptophan	
		Egg[2]	FAO[3]	Egg	FAO	Egg	FAO	Egg	FAO
Urban									
East	Low	80	102	57	94	72	92	76	118
	Middle	85	109	60	99	74	95	79	122
	High	88	113	63	104	75	96	80	123
North-east	Low	87	112	53	88	74	95	75	116
	Middle	89	114	61	100	75	96	79	122
	High	90	115	63	104	75	96	79	122
South	Low	71	91	59	97	70	89	77	119
	Middle	77	99	60	99	71	91	80	124
	High	87	112	62	102	75	96	82	127
Rural									
East	Low	78	100	55	90	74	94	72	112
	Middle	82	105	55	91	75	96	76	118
	High	86	111	54	88	76	97	77	120
North-east	Low	98	126	47	77	78	100	74	115
	Middle	97	124	52	85	79	101	78	120
	High	95	122	56	92	79	102	80	124
South	Low	76	97	55	91	72	92	74	115
	Middle	81	103	57	94	75	95	79	123
	High	86	111	58	96	76	98	80	124

[1]*Data from Jansen et al (1977)*
[2]*Amino acid Score Egg*

$$= \frac{mg\ AA_n/g\ N\ in\ dietary\ pattern}{mg\ AA_n/g\ N\ in\ whole\ egg} \times 100.$$

[3]*Amino acid score (FAO/WHO 1973)*

$$= \frac{mg\ AA_n/g\ N\ in\ dietary\ pattern}{mg\ AA_n/g\ N\ in\ 1973\ FAO\ pattern} \times 100.$$

SUMMARY

There is little doubt of the importance of protein quality
in animal nutrition. The evidence reviewed in this paper suggests that protein quality is important in human nutrition as
well, especially in relation to growth and lactation. Practical diets in many developing countries are marginal in protein
quality and calorie density. If such diets are increased in
calorie density, protein quality would appear to become even
more marginal.

Hegsted (1973) has pointed out that some widely used protein quality assays overestimate the quality of cereal diets.
He suggests that it is more dangerous to overestimate than underestimate the protein value of poor quality cereal diets.
Garza et al (1977) published evidence indicating that the 1973
FAO/WHO report (1973) underestimates the protein needs of
adult human subjects. Recently, Harper (1977) in reviewing
the U.S. dietary goal to reduce fat consumption to 30% of calories made the following significant comment:

"The recommendation to substitute cereal grains for meat
and eggs to accomplish this result would reduce the quality of
the protein in the diet and probably the amount of the protein. It is doubtful that this proposal would be of any concern in meeting nutrient needs of adults, but if families were
to shift the diets of young children toward cereal grains and
plant products on the basis of this recommendation being
adopted as a dietary goal, it would be a step toward accepting
the types of diets consumed in most poor countries of the
world where nutrition problems are of major concern. It would
be a long step backward for mankind if dietary goals designed
to alleviate chronic diseases in adults were to result in deterioration of the diets of young children."

Diets in most developing countries where cereals are the
major food staples currently have considerably lower levels of
fat and a higher proportion of protein of cereal origin than
would likely ever be achieved in the United States, even if
the "dietary goals" of the McGovern Committee were exceeded.
It would appear that, in considering nutrition improvement in
developing countries, improvement of the nutritional quality
of the diet, in protein, fat, vitamins and minerals, is still
a legitimate and worthy goal.

REFERENCES

Boongaart, J.; Baker, D. H.; Jensen, A. H. and Harmon, B. G. (1972). J. Animal Sci. 34: 408-410.

Bressani, R. (1977a). Evaluation of Proteins for Human Consumption. Bodwell, C. E. (Ed.) pp. 81-118, AVI Press, Westport, CT.

Bressani, R. (1977b). Evaluation of Proteins for Human Consumption. Bodwell, C. E. (Ed.) pp. 204-232, AVI Press, Westport, CT.

Bressani, R.; Alvarado, J. and Viteri, F. (1969). Arch Latinoamer. Nutr. 19: 129-132.

Edozien, J. C.; Khan, M. A. and Waslien, C. I. (1976). J. Nutrition 106: 312-328.

El Lozy, M.; Hegsted, D. M.; Herr, G. R.; Boutourline, E.; Test, G.; Ghamry, M. T.; Stare, F. J.; Kallal, Z.; Turki, M. and Hemaidan, N. (1975). Am. J. Clin. Nutr. 28: 1183-1188.

FAO/WHO Energy and Protein Requirements (1973). World Health Organ. Tech. Rep. Ser. 522, Geneva.

Fomon, S. J. (1959). Pediatrics 24:577-584.

Garza, C.; Scrimshaw, N. S. and Young, V. R. (1977). J. Nutrition 107: 335-352.

Gershoff, S. N.; McGandy, R. B.; Suttapreyasri, D.; Promkutkao, Nondasuta A.; Pisolyabutra, U.; Tantiwongse, P. and Viravaidhoya, V. (1977). Am. J. Clin. Nutri. 30: 1185-1195.

Graham, G. G. (1971). Amino Acid Fortification of Protein Foods. Scrimshaw, N. S. and Altschul, A. M. (Ed.). pp. 222-236 M.I.T. Press, Cambridge, MA.

Graham, G. G.; Baertl, J. M. and Placko, R. P. (1972). Agricultural and Food Chemistry 20: 506-508.

Graham, G. G.; MacLean, W. C., Jr. and Placko, R. P. (1976). J. Nutrition 106: 1307-1313.

Graham, G. G.; Morales, E.; Cordano, A. and Placko, R. P. (1971). Am. J. Clin. Nutr. 24: 200-206.

Graham, G. G.; Placko, R. P.; Acevedo, G.; Morales, E.; Cordano, A. (1969). Am. J. Clin. Nutr. 22: 1459-1468.

Harper, A. E. (1962). Am. J. Clin. Nutr. 11: 382-388.

Harper, A. E. (1977). J. Nutrition Education 9: 154-156.

Hegsted, D. M. (1973). Proteins in Human Nutrition. Porter, J. W. G. and Rolls, B. A. (Ed.) Academic Press, NY p. 366.

Hoffman, W. S. and McNeil, G. C. (1949). J. Nutrition 38: 331-337.

Jansen, G. R. (1974). New Protein Foods. Altschul, A. M. (Ed) Vol. 1A, pp. 39-120 Academic Press, New York.

Jansen, G. R. (1977). Evaluation of Proteins for Human Consumption. Bodwell, C. E. (Ed.) pp. 177-203 AVI Press, Westport, Ct.

Jansen, G. R.; Deuth, M. A.; Ward, G. M. and Johnson, D. E. (1975). Nutr. Rep. Int. 11: 525-536.

Jansen, G. R.; Jansen, N. B.; Shigetomi, C. T. and Harper, J. M. (1977). Am. J. Clin. Nutr. 30: 955-964.

Jansen, G. R. and Monte, W. C. (1977). J. Nutrition 107: 300-309.

Kies, C. and Fox, H. M. (1971). J. Food Sci. 36: 841-845.

Leung, P. M. B. and Rogers, Q. R. (1969). Life Sciences 8II: 1-9.

Longenecker, J. B. and Hause, N. L. (1959). Arch. Biochem Biophys 84: 46-55.

Lonnerdal, B.; Forsum, E.; Gehre-Medhim, M.; and Habraeus, L. (1976).

MacLean, W. C., Jr.; Placko, R. P. and Graham, G. G. (1977). J. Nutrition 107: 567-573.

Payne, P. R. (1975). Am. J. Clin. Nutr. 28: 281-286.

Reddy, V. and Gupta, C. P. (1974). Am. J. Clin. Nutr. 27: 122-124.

Sammonds, K. W. and Hegsted, D. M. (1973). Am. J. Clin. Nutr. 26: 30-40.

Scrimshaw, N. S. (1977). Nutrition Reviews 35: 321-337.

Scrimshaw, N. S.; Taylor, Y. and Young, V. R. (1973). Am. J. Clin. Nutr. 26: 965-972.

Wilcke, H. L. (Ed.). Improving the Nutrient Quality of Cereals. Report of Second Workshop on Breeding and Fortification. University of Colorado Conference Center, Boulder, CO, Sept. 12-17, 1976. Published by Agency for International Development, 1977.

Young, V. R.; Fajardo, L.; Murray, E.; Rand, W. M. and Scrimshaw, N. S. (1975). J. Nutrition 105: 534-542.

Young, V. R.; Tontisirin, K.; Ozalp, I.; Lakshmanan, F. and Scrimshaw, N. S. (1972). J. Nutrition 102: 1159-1170.

Zezulka, A. Y. and Calloway, D. H. (1976). J. Nutrition 106: 212-221.

DISCUSSION

Abernathy: You were concerned about the protein requirement for children. I would like to point out that the latest NRC statement on protein recommended allowance of 7- to 10-year-old children is 6% of calories, whereas for the adult over 51, it is more than 9%. If the NRC is right, we should be concerned with adults rather than children. If the NRC is wrong, we should change the NRC statement.

Jansen: There are other people here, particularly Dr. Scrimshaw, who have discussed the problems in trying to calculate protein-energy ratios. Some people have high protein-low energy, others low protein and high energy requirements. We know that the protein as a percentage of calories in breast milk is around 7%, but it is an extremely well-balanced and digestible protein. I think that it is difficult to observe 7- to 9-year-old children growing anywhere in the world on diets where the protein intake as a percentage of calories is in the neighborhood of 6-7%. As far as the adults are concerned, what you say might be true; there is a value in being concerned about the adults. However, if you can't make a case for the importance of protein quality for a growing child or a lactating mother, you are going to have trouble making it for anybody.

Abernathy: I'm not disagreeing with you, I'm just disagreeing with the NRC that says a generous protein allowance for the 7- to 10-year-old child is 6% of the calories of protein. Corn grits has 8.5 to 9.3%.

Jansen: I agree with you.

Harper: We should get this in perspective. The protein allowance of 1.2 gm/kg body wt/day for the child of 7-10 years is set as an allowance for protein, not as a percent of calories. When you calculate calorie-protein ratios, two things can change: protein needs and energy needs. The protein requirement of the adult, 51 years of age or older, is 67% of that of the child; the average calorie requirement is 42% of that of the child. The calorie requirement decreases much more with age than the protein requirement; therefore, the protein requirement of the adult, expressed as a percent of average calorie requirement, must be higher than that of the 7- to 10-year-old child. This just illustrates that requirements or allowances expressed as a percent of calories vary inversely with energy expenditure. A 30 kg child whose caloric expenditure is 1800 kcal would need 8% of calories as protein, a similar child whose caloric expenditure is 3000 kcal would need only 5% of calories as protein - but both need 1.2 gm of good quality protein/kg of body weight.

Scrimshaw: Well, it is interesting that taking the amount of egg or meat or beef protein that we find necessary to bring MIT students into nitrogen balance when energy needs are met and at energy intakes which are normal for them, the mean plus 2 standard deviations for both energy and energy intake as observed and protein intake as observed necessary for balance

comes out somewhere a little over 9%. As far as the children
ratios, Whitehead has emphasized quite strongly, and some others now are picking this up and supporting it, that the
FAO/WHO committee not only in '71 but consistently probably
made a mistake in taking as the criterion the average daily
increase in weight as simply 1/365 of the year's gain, because
children simply do not grow that way. They will grow 2 or 3
times that for a period of weeks or months and then maybe not
at all. Now, there is probably enough leeway for this
variation in a normal population; but as soon as you get into
a population where catch-up growth becomes a problem for
episodes of diarrhea or infection, then what Whitehead has
observed in the Zambia is that in the better-off families a
catch-up growth rate of 5 times can occur. Whereas in the
poorer families there may be little catch-up at all; and on
the metabolic ward where they can give a high protein and
energy intake, rates may go up as high as 18. Now, the
significance to this is that the protein intake goes up much
faster with catch-up growth than the energy intake; and if you
are going to allow for 3 times the average daily rather than
just the daily, then your percentage of protein calories
certainly has been increased.

HUMAN REQUIREMENTS FOR LYSINE
AND SULFUR-CONTAINING AMINO ACIDS

A. E. Harper

The literature on amino acid requirements of infants (Holt, et al 1961 and 1965; Fomon and Filer 1967), children (Holt, et al, 1965 and Nakagawa 1964) and human adults (Rose, 1957; Leverton, 1959; Hegsted, 1963; Munro, 1972; Harper, 1974) has been reviewed regularly. Irwin and Hegsted (1971) have summarized most of the information that is available on the subject; a comprehensive bibliography has been compiled by Ghadimi and Tejini (1975) and Williams et al (1974) have evaluated amino acid requirement studies critically. Despite the number of reviews, and the list is not exhaustive, there have been few experimental investigations of human requirements for amino acids. As a result, many questions concerning amino acid requirements remain unanswered, such as the rate of change of requirements with age, variation in requirements among individuals, and the extent to which requirements are influenced by changes in diet composition and energy intake and the physiological state of the subjects.

Human protein requirements have been investigated exhaustively (FAO/WHO, 1973; NAS/NRC 1974) and, since the protein requirement is a composite of the requirements for specific amino acids, inferences concerning amino acid requirements have been drawn from information about protein requirements. In fact, inferences of this type provide the most reliable information that is available about the amino acid requirements of infants, and serve as the basis for much of our knowledge about the rate of change of amino acid requirements between infancy and adulthood.

Although my task is to deal only with requirements for lysine and the sulfur-containing amino acids, it seems appropriate to deal with them in the context of amino acid requirements generally.

Table 1. Amino Acid Requirements of Infants

	Estimated Amino Acid Requirements		
		Based on Milk Protein Intake	
	Holt and Snyderman (12)	Williams et al (12)	FAO/WHO (13)
		mg/kg body wt/day	
Histidine	34	33	28 (39)[1]
Isoleucine	100	83	70
Leucine	150	135	161 (140)
Lysine	103	99	161 (99)
Met + cys	45 + cys	49	58 (50)
Phe + tyr	90 + tyr	141	125
Threonine	87	68	166 (66)
Tryptophan	22	21	17 (21)
Valine	105	92	93

[1]Figures in parentheses calculated from average values for milk composition (13).

REQUIREMENTS OF INFANTS

The two main sources of information about amino acid requirements of infants are the investigations of Holt, Snyderman and their associates (1961, 1965) who estimated minimum amounts of amino acids required to maintain a satisfactory rate of growth of infants consuming liquid formula diets containing free amino acids and those of Fomon, Filer and associates (1967) who estimated amino acid requirements from minimum amounts of human or cow's milk or soybean protein consumed by infants having satisfactory rates of growth. Values derived from both types of studies have been assessed by committees which have reviewed amino acid requirements as the basis for devising amino acid scoring patterns for nutritional evaluation of food proteins (FAO/WHO, 1973; NAS/NRC, 1974). Estimates of amino acid requirements are summarized in Table 1. Some of the discrepancies between the values selected by these committees have been discussed previously (Harper, 1977).

Lysine.--The lysine requirement of human infants consuming an amino acid formula, in which the lysine content was adjusted downward until it just failed to maintain a satisfactory rate of growth, was 103 mg/kg of body weight. This was the value for the infant with the highest requirement (Snyderman, 1974). Fomon (1974) estimated the lysine requirement of infants fed various formulas that contained cow's milk or soybean proteins to be 101 mg/100 kcal. Williams et al (1974) compiled the values from the various studies done by Snyderman et al (1974) and Fomon et al (1973) and recalculated the amino acid intakes from the information on formula consumption published by Fomon et al (1973) using USDA amino acid composition values for the proteins in the formulas. They concluded that an appropriate value for the lysine requirement of infants 4 to 6 months of age was 99 mg/kg of body weight/day. All of these values are clearly in agreement. They are also in agreement with the value of 102 mg/kg given in a more recent report by Snyderman (1974).

Methionine.--The requirement of the infant for methionine was studied by Holt, Snyderman et al (1965) only in the presence of a surplus of cystine (about 80 mg/kg of body weight), so it is not possible to obtain an accurate value for the total sulfur-containing amino acid (methionine + cystine) requirement from their investigations. When proteins are the source of amino acids, methionine and cystine are consumed together. Figures for methionine + cystine requirement can be estimated from values for protein intakes of infants fed various milk and soybean formulas. Fomon (1974) estimated, from such calculations, that the methionine + cystine requirement of the infant was 47 mg/100 kcal. Williams et al (1974) after reviewing the available literature estimated the methionine cystine requirement of the infant 4 to 6 months of age to be 49 mg/kg of body weight/day. These values are in good agreement but are much lower than the FAO/WHO (1973) estimate of 58 mg/kg of body weight/day.

AMINO ACID REQUIREMENTS OF INFANTS AND AMINO ACID COMPOSITION OF COW'S MILK

Both FAO/WHO (1973) and Williams et al (1974) have devised amino acid scoring patterns for evaluating food proteins based on infant requirements for amino acids (Table 2). The basis for these scoring patterns is that protein of high quality, when fed in a quantity that meets the nitrogen requirement of the infant, will also meet all of the requirements for indispensable amino acids. The way in which the scoring patterns

Table 2. Amino Acid Scoring Patterns

	NAS/NRC	FAO/WHO
	mg/gm protein	*mg/gm protein*
Histidine	17	--
Isoleucine	42	40
Leucine	70	70
Lysine	51	55
Met + cys	26	35
Phe + tyr	73	60
Threonine	35	40
Tryptophan	11	10
Valine	48	50

have been calculated is shown in Table 3. The major difference between the scoring patterns devised by FAO/WHO (1973)
and Williams et al (1974) is the lower value for methionine +
cystine accepted by Williams et al (1974).

It is evident from comparisons between amino acid requirements of infants determined by Holt and Snyderman (1961) using
formulas that contained free amino acids and requirements estimated from amino acid intakes of infants consuming various
formula diets (Williams, et al 1974) that the lowest estimates
for most amino acids have been obtained by the latter method.
Values obtained for several amino acids by the two methods are
in close agreement. As one would expect from this information, there is close correspondence between the cow's milk
amino acid pattern, expressed as mg of amino acid/gm of protein, and amino acid requirements of infants expressed in the
same terms (Figure 1). The lysine requirement has been established from intakes of lysine from human milk and amino acid
diets and the methionine + cystine requirement from intakes of
these amino acids from cow's milk formulas.

Arroyave (1974) has used the FAO/WHO scoring pattern to
predict the quantities of milk and corn-bean mixtures that
would be required to meet the amino acid requirements of 2-
year old children, on the assumption that requirements for
amino acids fall at least as rapidly with age as does the total nitrogen requirement. He predicted that 0.9 gm of milk
protein/kg of body weight/day would meet the requirement of
these children for methionine + cystine, the limiting amino
acid in cow's milk, and that 1.33 gm of the corn-bean mixture
would meet the lysine requirement, the limiting amino acid in

Table 3.   Basis for Amino Acid Scoring Patterns

Lysine requirement
   of infant = 99 mg/kg/ body wt/day

Protein requirement
   of infant = 2 gm/kg body wt/day

Therefore, a product that contains
   50 mg of lysine/gm of protein
will meet the lysine requirement of
the infant when it is fed in a
quantity that will meet the nitrogen
requirement.

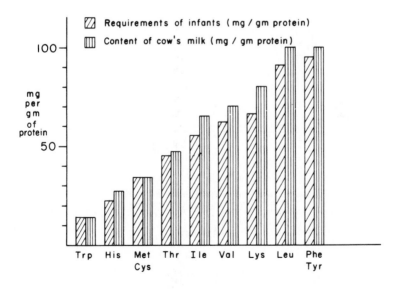

FIGURE 1.   Amino Acid Patterns of Infant Requirements and
Milk Protein.

the mixture, and that these amounts of both proteins would
meet the nitrogen requirement. In reviewing, retrospectively,
studies in which these proteins had been tested as nitrogen
sources for children, the predictions were in good agreement
with the results obtained. This would indicate that amino acid requirements per gram of protein are no higher, and are
probably lower, than the amounts per gm of protein provided by
milk proteins.

Because amino acid requirements fall sharply with age,
amino acid scoring patterns would be expected to underestimate
the nutritional value or quality of a food protein except for
very young infants. Snyderman (1974) has shown that the requirement of the infant for phenylalanine falls sharply during
the first year of life. Her results would suggest that amino
acid requirements fall within 18 months to less than one-third
of the value for the newborn (Figure 2). Protein requirements
are considered to fall by only about 50% during this time period. This implies that protein quality is of much less significance after age one than before. This becomes clear by
looking further at the amino acid requirements of children and
adults.

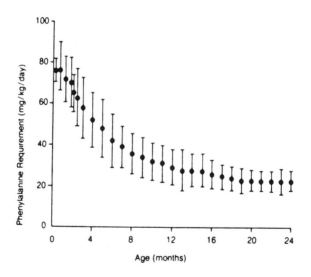

FIGURE 2. The phenylalanine requirement of the phenylke-
tonuric infant related to age; average of 50 infants and one
standard deviation above and below the average (Snyderman,
1974).

REQUIREMENTS OF CHILDREN

Lysine.--Nakagawa and associates (1964) have estimated the amino acid requirements of 12 to 14 year old boys. They used rather wide increments of amino acids in their nitrogen balance study. The requirement for lysine could be estimated only as falling between 45 and 60 mg/kg of body weight/day. Williams et al (1974) estimated by regression from a plot of the individual values, that the requirement was probably close to 45 mg/kg. This is a little less than half the requirement of the infant, and is above the value predicted from the rate of fall in the requirements of some of the other amino acids (Snyderman, 1974).

Sulfur-Containing Amino Acids.--The value for the methionine requirement for boys was determined by Nakagawa et al (1964), in the absence of cystine, to fall between 14 and 27 mg/kg of body weight/day. The estimate made from the individual values by Williams et al (1974) was about 22 mg/kg. This is just under half of the infant requirement and again the fall is less rapid than would be predicted from the fall in the requirements for some of the other amino acids. The FAO/WHO Committee (1973) took Nakagawa's highest value of 27 mg/kg/day as the value for the scoring pattern. This resulted in a value for methionine + cystine per gm of protein that was higher than the requirement of the infant.

REQUIREMENTS OF ADULTS

Lysine.--Lysine requirements for men and for women have been estimated as 11.4 and 8.6 mg/kg of body weight/day, respectively (Williams et al, 1974). Hegsted (1963), on the basis of regression analyses of the values for women, estimated lysine requirement to be 9.4 mg/kg of body weight/day (Table 4). Williams et al (1974) accepted this as the most reliable estimate of the adult lysine requirement. Subsequent observations by Young et al (1972) who investigated nitrogen balance and the response of plasma lysine concentration in young men fed graded levels of lysine in an amino acid diet would indicate that the lysine requirements of adults do not exceed this value and may actually be below it. In short-term nitrogen balance studies Fisher et al (1969) obtained a value of only about 1 mg/kg of body weight for young women. This low value for the adult requirement for lysine is not readily explained. If the lysine requirement of adults is much below 8 to 9 mg/kg of body weight/day, it is certainly difficult to account for

Table 4. Amino Acid Requirements of Adults

| | Estimated Amino Acid Requirement | | |
| | Rose (5) | Leverton (6) | Hegsted (7) |
		mg/kg of body wt/day	
Histidine	?	?	?
Isoleucine	10.1	7.8	9.5
Leucine	15.7	10.7	12.5
Lysine	11.4	8.6	9.4
Met + cys	14.4	12.1	12.1
Phe + tyr	15.7	12.1	12.1
Threonine	7.1	5.3	6.5
Tryptophan	3.6	2.8	2.9

observations that lysine fortification will improve the efficiency of utilization of white flour by adult human subjects.
It is also difficult to account for the large amount of nitrogen (7.4 gm/day) required from flour to achieve N-equilibrium
in adults unless lysine is the limiting amino acid (FAO/WHO,
1973; NAS/NRC, 1974). If lysine requirement were uniquely
low, 5 to 6 gm of flour protein rather than 7.4 gm would be
expected to meet the nitrogen requirement.

Sulfur-Containing Amino Acids.--Total sulfur-containing
amino acid requirements of adults have been estimated for men
and women, respectively, as 14.4 and 12.1 mg/kg of body
weight/day (Williams, et al, 1974). Hegsted (1963) on the basis of regression analysis of the values for women has estimated the requirement to be 12.1 mg/kg. Williams et al (1974)
accepted this as the most reliable estimate of the methionine
+ cystine requirement. In a reinvestigation of the methionine
cystine requirement Zezulka and Calloway (1976) noted that no
man in their study required more than 12 mg of methionine +
cystine/kg of body weight/day and most required much less.

Hegsted (1963) has emphasized that there is considerable
variability among individuals with respect to amino acid requirements. However, if the range of amino acid requirements
is assumed to be plus or minus three times a hypothetical coefficient of variation of 15% from the midpoint value, the
range found experimentally is essentially that predicted
(Harper, 1977; 1977). Such a range is typical of biological
variables. This, together with the predictability observed by
Arroyave (1974) using a scoring pattern based on amino acid
requirements, indicates that values for amino acid require-

ments, despite their shortcomings, are reasonable and should serve as a satisfactory guide to human needs.

It is of some interest to look at changes in amino acid requirements with age and the ratios of infant to adult requirements (Table 5). The nature of the change was pointed out by Holt et al (1960) many years ago. It is evident that requirements for most amino acids fall more rapidly with increasing age than does the requirement for nitrogen. They fall by maturity to about one-ninth of the infant value whereas the nitrogen requirement falls to about one-third of the infant value. Lysine requirement fits the pattern, but interestingly methionine + cystine does not. It falls only to about one-quarter of the infant requirements if present estimates of methionine + cystine requirements are correct. There is some evidence from studies in which proteins and nonspecific nitrogen have been fed to human subjects by Swendseid et al (1959) that this requirement may be overestimated. However, the possibility also exists that the requirement of adults for methionine + cystine is proportionately higher than for other amino acids. Nevertheless, the methionine + cystine requirement of adults is low enough so that there is little likeli-

Table 5. Comparison of NAS/NRC Estimated Amino Acid Requirements of Infants and Adults

	Infant (4-6 mo.)	Adult	Infant/Adult
	mg/kg of body wt/day		
Protein	1500	470	3.2
Histidine	33	?	--
Isoleucine	83	9.5	8.7
Leucine	135	12.5	10.8
Lysine	99	9.4	10.5
Met + cys	49	12.1	4.1
Phe + tyr	141	12.1	11.6
Threonine	68	6.5	10.5
Tryptophan	21	2.9	7.2
Valine	92	10.7	8.6
Total minus histidine	681	76	Av. 9.3

hood of a problem being encountered in meeting it from ordinary mixed diets or proteins of even reasonable quality, when the nitrogen requirement is met.

Where, then, do we come out?

Despite the criticisms that are made about the inadequacy of our knowledge of amino acid requirements, amino acid scoring patterns based on these requirement values permit prediction, with a reasonable degree of confidence, of the effectiveness of *highly digestible proteins* in meeting human protein needs.

In my judgment, the FAO/WHO (1973) scoring pattern overestimates the need for sulfur-containing amino acids. The scoring pattern proposed by Williams et al (1974) is similar in most respects but the lower value for sulfur-containing amino acids approximates more closely the sulfur-containing amino acid requirement of the infant (Fomon, 1973). These scoring patterns are based on the requirements of infants for whom protein quality is most critical. Amino acid requirements decrease with increasing age so proteins that meet the amino acid needs of the infant should be more than adequate for older age groups.

Finally, our major concern is with prediction of the efficiency of utilization of food proteins by human beings and with prediction of the extent to which the various proteins that are present in a mixed diet will complement each other in meeting overall nitrogen and amino acid needs. None of the standard methods of estimating protein quality by bioassay procedures are useful for this purpose (NAS/NRC, 1974). However, nitrogen requirements and amino acid requirements are determined under conditions in which efficiency of utilization is part of the measurement. Therefore, when amino acid requirements serve as the basis for an amino acid scoring pattern the amount of an amino acid that is needed to meet the score value includes an allowance for the fall in efficiency of utilization that occurs as amino acid intake approaches the requirement value (FAO/WHO, 1973). Use of the scoring pattern is the only way in which the extent to which proteins supplement or complement each other can be predicted. It is the only rational way of assessing the protein quality of human foods and diets.

The major shortcoming of this procedure is that amino acids are not always released completely during digestion. The chemical analysis on which the scoring procedure depends may overestimate the nutritional value of a protein if some amino acids are only partly available. Progress in measuring biological availability of amino acids is less than is desirable. It would, therefore, seem necessary, if there is a need and desire for a dependable method of predicting efficiency of protein utilization, and for a satisfactory method for labeling foodstuffs for protein content, that those concerned should focus attention on the need for effective and reliable methods for measuring bioavailability of amino acids. Once such measurements can be obtained routinely, it will be possible to include a correction for availability in comparing the amino acid composition of a protein with the standard scoring pattern.

REFERENCES

Arroyave, G. (1974). Nutrients in Processed Foods - Proteins (P. L. White and D. C. Fletcher, eds.). Publishing Sciences Group, Inc., Acton, MA, pp. 15-28.

Fisher, H.; Bush, M. K.; and Griminger, P. (1969). Am. J. Clin. Nutr. 22: 1190-1196.

Fomon, S. J. (1974). Infant Nutrition, Second Edition. W. B. Saunders Co., Philadelphia.

Fomon, S. J. and Filer, L. J. Jr. (1967). Amino Acid Metabolism and Genetic Variation (W. L. Nyhan, ed.). McGraw-Hill, New York, pp. 141-143.

Fomon, S. J.; Thomas, L. N.; Filer, L. J. Jr.; Anderson, T. A.; and Bergmann, K. E. (1973). Acta Paediat. Scand. 62: 33-45.

Food and Agriculture Organization/World Health Organization (1973). WHO Technical Report Series No. 522, FAO Nutrition Meetings Report Series No. 52. FAO/WHO, Geneva.

Ghadimi, H. and Tejani, A. (1975). Total Parenteral Nutrition - Premises and Promises (Ghadimi, H. ed.). John Wiley & Sons, New York, pp. 213-230.

Harper, A. E. (1974). Protein Nutrition (H. Brown, ed.). Thomas, C. C., Springfield, IL, pp. 130-179.

Harper, A. E. (1977). Clinical Nutrition Update - Amino Acids (H. L. Greene, M. A. Holliday, and H. N. Munro, eds.). American Medical Association, Chicago, IL, pp. 58-65).

Harper, A. E. (1977). Food Proteins (J. R. Whitaker and S. R. Tannenbaum, eds.). AVI Publishing Co., Inc., Westport, CT, Chap. 15, pp. 363-386).

Hegsted, D. M. (1963). Fed. Proc. 22: 1424-1430.

Holt, L. E., Jr.; Gyorgy, P.; Pratt, E. L.; Snyderman, S. E.; and Wallace, W. M. (1960). Protein and Amino Acid Requirements in Early Life. New York University Press.

Holt, L. E., Jr., and Snyderman, S. E. (1961). J. Am. Med. Assoc. 175: 100-103.

Holt, L. E., Jr., and Snyderman, S. E. (1965). Nutr. Abstr. Rev. 35: 1-13.

Irwin, M. I. and Hegsted, D. M. (1971). J. Nutr. 101: 539-566.

Leverton, R. M. (1959). Protein and Amino Acid Nutrition (A. A. Albanese, ed.). Academic Press, New York, pp. 477-506.

Munro, H. N. (1972). Parenteral Nutrition (A. W. Wilkinson, ed.). Churchill-Livingston, London, p. 34.

Nakagawa, I.; Takahashi, T.; Suzuki, T.; and Kobayashi, K. (1964). J. Nutr. 83: 115-118.

National Academy of Sciences/National Research Council (1974). Improvement of Protein Nutriture. NAS/NRC, Washington, DC, pp. 45-47.

National Academy of Sciences/National Research Council (1974). Recommended Dietary Allowances, Eighth Edition. NAS/ NRC, Washington, DC.

Rose, W. C. (1957). Nutr. Abstr. Rev. 27: 631-647.

Snyderman, S. E. (1974). Heritable Disorders of Amino Acid Metabolism: Patterns of Clinical Expression and Genetic Variation (W. L. Nyhan, ed). John Wiley & Sons, New York, Chap. 29, pp. 641-651.

Swenseid, M. E.; Feeley, R. J.; Harris, C. L.; and Tuttle, S. G. (1959). J. Nutr. 68: 203-211.

Williams, H. H.; Harper, A. E.; Hegsted, D. M.; Arroyave, G.; and Holt, L. E. Jr., (1974). Improvement of Protein Nutriture, National Academy of Sciences, Washington, DC., pp. 23-63.

Young, V. R.; Tontisirin, K.; Ozalp, I.; Lakshmanan, F.; and Schrimshaw N. S., (1972). J. Nutr. 102: 1159-1169.

Zezulka, A. Y. and Calloway D. H. (1976). J. Nutr. 106: 212-221.

DISCUSSION

Abernathy: I would like to comment on the requirements for the children as stated by Nakagawa et al. Their requirements for girls are calculated on the basis of a nitrogen retention of about 0.38 of a gram per day of nitrogen retention. This retention covers growth but not sweat losses, so the values probably are low on that account. They are also calculated on the basis of low energy intake, which has the opposite effect. This probably cancels out the errors in estimations. We calculated amino acid needs, based on the nitrogen balance studies we've had from 7- to 9-year-old girls, and were generally a little lower; but we also calculated on the basis of 0.3 g, and so that makes ours probably a little low.

Their energy intakes were low and purified amino acids were used, which probaby could account for the amino acid requirements.

Bressani: There is something that has always worried me about the estimates of amino acid requirements, and that is that they are derived from studies either using mixtures of amino acids or proteins that are fed in such a way that all the amino acids studied were presented to the individuals simultaneously. Now, when we think in terms of real diets, I think that all of us will agree that we don't have this balance of amino acids in every meal that we take; and the same thing, of course, is true for when we are thinking in terms of using different protein sources to complement each other. I wonder if you would like to comment about the implications of not ingesting all these amino acids simultaneously.

Harper: We certainly know a little from the animal studies that have been done. If one amino acid is deleted completely from the diet and supplied a few hours later, for many amino acids there is little or no response at all. For lysine a response is still observed if it is provided several hours later. It is impossible for me to extrapolate from the extreme for which we do have figures to the other for which we don't. At the same time, I think one may have to allow for the fact that we have periods of lesser nitrogen retention, and in adults negative retention, for a period counterbalanced subsequently by periods of high positive retention. So there's a tendency for the negative and positive periods to balance out as long as overall intake is adequate over time. Secondly, even though we may not get exactly the requirement in a meal, the stomach is an excellent regulator of the rate of flow of digestion products to the rest of the body, so that time lag before the rest of the requirement is met may not be great. Periods without a particular amino acid are unlikely to occur with human subjects. In most of the animal studies where a very severe effect has been observed, diets have been completely devoid of an amino acid.

Scrimshaw: You might wish to comment on two points. One is that, so far as I know, efforts to add up the requirements for individual essential amino acids and feed them back have generally been unsuccessful; it has required a higher amount. The situation appears analogous to that for total protein, where the tendency has been to give high energy intakes and to feed at levels at which metabolic adaptation is extreme. The question arises as to whether the high energy intakes supplied by Rose and Leverton and co-workers might not be one factor in this equation, and the other the fact that their subjects were adapted at the low amino acid intakes to an extent that is impossible at requirement levels. Of course, the amino acid

proportions could still yield the satisfactory results that
you indicate at just a little higher concentration of total
protein.

The second question that is brought to mind is your notic-
ing out-liers in studies of amino acid requiremnts; and again,
by analogy, with the problem of studying total protein requirements. In every metabolic balance study involving any
significant number of subjects, there is likely to be an individual who behaves "improperly." Unfortunately, it is often
assumed that this is due to a technical error of some kind,
and the data are discarded. Hence, these out-liers or devi-
ants usually do not get reported. Calloway, too, has described an individual who was fed increasing amounts of protein, yet showed no increase in nitrogen retention. We see
such cases in studies with milk or beef or other protein
sources. I wonder how many out-liers there are for individual
essential amino acids, and also for protein requirements.

Harper: I don't know that I have answers for both of
these questions. Certainly, we have been concerned about how
low the amino acid requirements of adults are. I think every
person who has reviewed the amino acid requirements of adults
is struck by the fact that the requirements for essential amino acids represent less than 20% of the total protein requirement. When one looks at the studies, they are done under conditions in which you might expect highly efficient utilization
of that amino acid. As a rule there were surpluses of all of
the amino acids except one. One might expect a mass action
type of effect such that the component of the system that is
present in least amount would be used with the greatest efficiency. I estimated once that there could be an increase of
30% in efficiency under these conditions. I'm still not sure
that may not be right compared to conditions in which requirements are determined with everything in balance.

On the other hand, I'm struck with the fact that requirements of children determined by Snyderman et al with the concentration of one amino acid greatly reduced, were fairly
close to those obtained in studies done with protein by Fomon.
So there is still a question in my mind as to how much that
increased efficiency under conditions in which many of these
amino acid studies have been done is a factor.

The other question about the out-liers: I looked at all
of the figures for amino acid requirements (Hegsted has done
this, too,) and plotted most of them. There are relatively
few in the amino acid requirement studies on adults who fall
outside the statistically predicted range. But there are
some. One always has a question as to whether an incipient
infection or some unique physiological problem has occurred at

the time the studies are done. One out of a group of 8 or 10 individuals is not unusual, even in animal studies. I don't have the answer as to whether these are people with unique requirements who represent a unique population or not. We haven't much evidence of this occurring with amino acids, although it does for some of the vitamins. I am inclined to think that a genetic defect that led to an unusually high amino acid requirement would be lethal during early development.

Satterlee: Dr. Harper, during the isolation of a protein there is the possibility of oxidizing some of the sulfur-containing amino acids. Are there any data available as to just how available the damaged sulfur-containing amino acids or the methionine sulfoxide and/or cysteic acid would be to the human?

Harper: I don't have figures on that specifically. My impression is that they are not highly available to mammalian species. Fortunately, a good many of these studies were done with milk proteins that were not highly processed. Some of them were done with soybean proteins, and the analyses I think were for methionine and cystine specifically.

Bodwell: I believe actually there are some papers in which it has been shown that methionine sulfoxide is probably available to the rat. The one who should really be addressing this question is Dr. Zezulka. She's done some of this with humans.

Zezulka: I think Dr. Bodwell is referring to the excretion of methionine sulfoxide we found when we fed D-methionine to humans, which is what I reported at FASEB meetings in April, 1978. From what I can determine from the literature, L-methionine sulfoxide in the mature rat appears to be utilized at least 70%, but there is some question about how much is available to the immature animal.

Simpson: Dr. Harper, we generally lump cystine and methionine together. Some of the work on infused amino acids would suggest that there might be a specific requirement for cystine aside from its formation from methionine. Would you like to comment on that?

Harper: There certainly has been evidence that cystine may be an essential amino acid for the premature, low birth weight infant. There's a sharp fall in cystine concentration in blood of premature infants fed formulas containing no cystine. As far as I know, there isn't evidence of a specific

requirement for cystine by older age groups, although an interesting point is that Rose had a somewhat lower requirement
for total sulfurcontaining amino acids with a combination of
methionine and cystine than with methionine alone.

Zezulka: Again, we did attempt in our sulfur-amino acid
diet study in young adults that we did about a year ago to
compare L-methionine only with an equivalent amount L-
methionine and cystine; and as far as the six subjects we were
looking at were concerned, there wasn't any significant trend
showing that when cystine was contributing about 30% of the
methionine requirements there was any increase in the nitrogen
balance; but that is the limited data that I have.

EFFECT OF SOY PROTEIN ON TRACE MINERAL AVAILABILITY[1]

Boyd L. O'Dell

INTRODUCTION

With the increasing use of soybean protein in the human
diet and its substitution for animal protein, the importance
of assessing its nutritional properties is evident. Assessment of its contribution to trace element nutrition involves
not only estimation of trace element content but also evaluation of bioavailability. Bioavailability is defined here as
that proportion of a chemically determined nutrient which is
absorbed and utilized. Such an assay must involve an animal
and finally man himself.

I.   *Bioavailability and Its Evaluation*

Absorption and utilization of any given element depends
on such intrinsic factors as species and the physiological
state of the animal, but this discussion will be restricted to
the extrinsic or dietary factors that affect bioavailability.
Extrinsic factors that may affect bioavailability include:
   a) Chemical nature, or speciation of the element.
   b) Competitive antagonism among closely related ele-
       ments.
   c) Adsorption on large surfaces of insoluble compounds.
   d) Chelate or complex formation; this effect may be
       detrimental or beneficial.
   e) Microbial flora in the gut.
This discussion is further restricted to those factors associ-

[1]Contribution of the Missouri Agricultural Experiment
Station, Journal Series No. 8144. Supported in part by Public
Health Service Grant HL11614.

ated with soybean protein which may influence bioavailability.
For a more general treatment of the subject see an earlier review (O'Dell, 1972).

Various bioassays have been used to measure bioavailability of trace elements, but only iron and zinc have received major attention. In human nutrition, mineral balance
or net retention of an element is commonly employed, except
for iron, where the use of isotopes is more popular (Sharpe et
al, 1950). The regeneration of hemoglobin in iron-depleted
animals in also used (Fritz et al, 1970). For the evaluation
of zinc bioavailability, growth rate of animals fed limiting
levels of zinc (O'Dell, 1969 and O'Dell et al, 1972a) and total femur zinc (Momcilovic et al, 1976) have been employed.
Forbes and Parker (1977) used slope ratios of both gain and
total femur zinc in rats to evaluate zinc availability in full
fat soy flour. Their results are shown in Figures 1 and 2.
The weight gain data give a slope ratio of 54% compared to 34%
for the femur zinc data. Which is the more reliable value is
a moot question, but theoretically it would seem more valid to
consider zinc accretion during the assay than total zinc.
Such a treatment of the data gives a higher availability for
the soy protein and one in better agreement with the growth
data. In addition to the methods described, both intrinsical-
ly (Welch et al, 1974) and extrinsically (Evans and Johnson,
1977) labeled zinc have been used to evaluate zinc absorption.

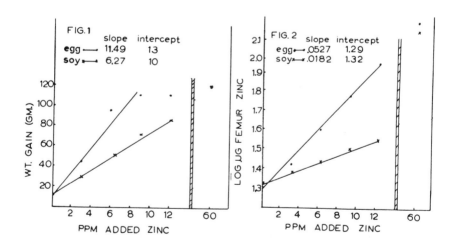

FIGURES 1 & 2. Use of slope ratios of weight and total
femur zinc to evaluate bioavailability of zinc in whole fat
soy flour compared to zinc carbonate added to egg white
protein. Taken from Forbes and Parker (1977).

II. Composition of Soybeans and Products

The major soybean products used for human food are defatted soy flour, protein concentrates and protein isolates. The concentrates are prepared by leaching with alcohol, acid or water to remove some of the carbohydrates. The protein isolates consist of the water soluble proteins and are nearly free of carbohydrate. Selected values of the gross compositions are shown in Table 1. For the purpose of this discussion the most important constituents are crude fiber and phosphorus. Calcium content is of secondary interest. Approximately 70% of the total phosphorus in soybeans and soybean protein products occurs in the form of phytate, inositol hexaphosphate. It should be pointed out that phytate could not occur, as is commonly stated, solely as phytin, Ca_5 Mg Phytate, because there is not sufficient calcium to account for this compound. This is certainly the case for the isolated protein which contains low concentrations of both Ca and Mg. Even whole soybeans contain less than 3 g atoms of Ca per mole of phytate.

Soybeans contain a complex mixture of carbohydrates which make up about one third of the mass. Of this about 5% may be classed as crude fiber. However, isolated soy protein contains no fiber.

Soybean trace elements of nutritional, or possible nutritional, significance are shown in Table 2. Of the 15 elements listed, there is little doubt about the nutritional significance of the first nine. The significance of some of the others is now under study. Nevertheless, from a practical standpoint Fe and Zn are of primary concern in human nutrition and the only ones for which appreciable bioavailability data have accrued.

Since phytate is the constituent of seeds of greatest concern relative to bioavailability of the trace metal cations, it is of interest to compare the phytate concentration in various products. The phytate content of selected seeds and plant proteins are shown in Table 3. Sesame and rapeseed meals have the highest phytate concentration with soybeans ranking higher than the cereal grains. In the case of corn, essentially all of the phytate is in the germ, while 87% of the phytate in wheat is in the aleurone layer and 80% of that in rice is in the pericarp (O'Dell et al, 1927b). Phytate is usually determined by precipitation with ferric ion in acid solutions. This reagent will precipitate inositol phosphate esters containing less than six phosphates. However, only the hexaphosphate of inositol was detected in mature corn, wheat, rice, sesame and soybean (Boland et al, 1975).

Table 1. Major Constituents of Soybeans and Soybean Products (Dry Basis)

	Whole Soybean[1]	Soy Flour[2]	Soy Concentrate[3]	Isolated Protein[2,3]
	%	%	%	%
Protein	42.9	51.6	71.6	98.7
Fat	19.6	0.93	0.37	—
Ash	5.0	—	6.5	3.2
Crude Fiber	5.5	2.8	3.8	0.2
NFE (Difference)	27.0	—	17.7	—
K	1.7	2.3	2.34	0.096
Ca	0.3	0.23	0.25	0.18
Mg	0.3	0.28	0.29	0.038
P	0.7	0.75	0.66	0.76
Phytate P	0.5	—	—	0.53

[1]Cartter and Hopper (1942).
[2]Lo (1978).
[3]Meyer (1978).

Table 2. Trace Element Content of Soybeans and Isolated Soy Protein[1]

Element	Defatted Soybeans[2]	Soy Protein[3]
	ppm	ppm
Fe	137	160
Zn	52	40
Cu	20	12
Mn	38	17
Cr	0.35	1
Mo	2	3
I	0.84	—
F	1.9	—
Se[4]	0.065	0.137
Si	140	—
Ni	6	2.5
V	0.04	10
As	0.04	—
Li	0.1	0.023[5]
Co	0.24	1

[1]Selected values for elements of nutritional or possible nutritional significance, dry basis.
[2]Osborn (1977) except Se.
[3]Lo (1978) except Se and Li.
[4]Se from Ferretti and Levander (1976).
[5]Patt and Pickett, personal communication.

Table 3. *Phytate Content of Selected Seeds and Proteins*[1]

Product	Phytate P
	%
Corn, commercial	0.25
Corn, high lysine	0.28
Wheat, soft	0.32
Rice, brown	0.25
Sesame meal, defatted	1.46
Rapeseed flour[2]	1.12
Pea seeds, mature[3]	0.33
Beans (Phaseolus vulgaris)[4]	0.18-0.39
Soybeans (15 varieties)[5]	0.28-0.41
Soybean meal, comm.	0.40
Soybean flakes, defatted	0.43
Isolated soy protein, comm.	0.43

[1]*Boland et al (1975) except as noted.*
[2]*Anderson et al (1976).*
[3]*Welch et al (1974).*
[4]*Lolas and Markakis (1975).*
[5]*Lolas et al (1976).*

III. *Bioavailability of Zn, Fe, Cu, Mn and Mg in Soybean Protein*

Zinc. The first observations that soybean protein affects trace element availability arose from zinc requirement studies. Chicks fed soybean protein had a higher zinc requirement than those fed casein and gelatin (O'Dell and Savage, 1957). It was soon shown that either autoclaving the soy isolate or addition of ethylenediaminetetracetate (EDTA) reduced the zinc requirement of turkey poults, pigs and rats fed diets based on soy protein (Kratzer et al, 1962; Smith et al, 1962; Forbes, 1964). In the latter study a strong inter-action was observed between calcium and soy protein as regards zinc requirement. Observations similar to those made with soy protein have been made with sesame meal (Lease et al, 1960) and rapeseed flour (Anderson et al, 1976). Although less of the zinc in soy protein is absorbed and utilized than of that in animal protein, this fact may not be of practical impor-tance when there is only partial substitution of soy for meat protein in the human diet. Greger et al (1978) substituted rehydrated defatted soy (Textured Vegetable Protein), for 30% of the meat in the lunch of adolescent girls and found no ef-fect on fecal or urinary zinc loss.

A few attempts have been made to quantitatively assess the biological availability of zinc in plant and animal food-stuff. Values relative to zinc salts, such as $ZnCO_3$ or $ZnSO_4$, are shown in Table 4. In general the zinc in animal products is more available than that in plant products, soy-bean zinc being 50-65% available. It has been postulated that phytate accounts for the low bioavailability of zinc in seed proteins (O'Dell and Savage, 1960). The effect of phytate will be discussed below.

Iron. It is generally concluded that the iron, par-ticularly heme iron, in animal foods has a higher bioavail-ability than that in plant foods (Bowering et al, 1976). Based on this generalization one would expect the iron in soy-bean products to be poorly utilized. In confirmation of this concept Fitch, et al (1964) observed a lower absorption of iron in Rhesus monkeys fed soybean protein than those fed casein. However, bioavailability assessment by hemoglobin repletion in the rat suggests a relatively high availability. Steinke and Hopkins (1978) observed an average of 61% bioavailability of the iron in isolated soy protein.

Similar high iron availability was observed in the grow-ing chick (Davis et al, 1962). Whereas the zinc in isolated soy protein was poorly utilized unless EDTA was added, there was no effect from adding EDTA when iron was limiting. Con-sidering the fact that ferric iron forms highly insoluble com-plexes with phytate, one might expect a low availability of iron in foods rich in phytate. Soybean protein is relatively high in both iron and phytate but the chemical form of the iron is unknown. Morris and Ellis (1976) have characterized a major iron component of wheat bran as monoferric phytate and showed that this soluble iron salt is as available as ferrous ammonium sulfate. Although phytate may have an effect on iron availability, it seems clear that phytate is not the only fac-tor involved. Welch and van Campen (1975) labeled soybeans with ^{59}Fe and administered single oral doses of immature and

Table 4. Bioavailability of Zinc in Feed and Foodstuffs

Product	Assay Animal	Method	Value
			%
Milk based infant formula	Rat	Slope ratio-femur	86[1]
Soy based infant formula	Rat	Slope ratio-femur	67[1]
Whole fat soy flour	Rat	Slope ratio-femur	34[2]
Whole fat soy flour	Rat	Slope ratio-gain	54[2]
Soybean meal	Chick	Growth rate	67[3]
Sesame meal	Chick	Growth rate	59[3]
Corn	Chick-rat	Growth rate	63,57[3]
Wheat	Chick-rat	Growth rate	59,38[3]
Rice	Chick-rat	Growth rate	62,39[3]
Egg yolk	Chick-rat	Growth rate	79,76[3]
Non-fat milk	Chick-rat	Growth rate	82,79[3]

[1]Momcilovic et al (1976).
[2]Forbes et al (1977).
[3]O'Dell et al (1972).

mature seeds to rats. Approximately 32% of the iron in the
immature seeds, which contained 0.61% phytic acid, was absorbed compared to approximately 52% in the mature seeds,
which contained 1.71% phytic acid. Apparently the immature
seeds contain a factor other than phytate which impaired iron
absorption. However, one must recognize that absorption of an
ion from a single dose may be different from that in a diet
consumed ad libitum.

Copper, Manganese and Magnesium. There is little information in the literature relative to the bioavailability of
these elements. As mentioned above, addition of EDTA to a soy
protein based diet increases the availability of zinc. Davis,
et al (1962) also observed that the addition of 0.07% EDTA im-

proved the response of chicks fed soy protein when either copper or manganese was limiting. EDTA had no effect when these
elements were adequate. On this basis it was concluded that
soybean protein binds Zn, Cu and Mn and makes them less biologically available.

Strictly speaking, magnesium is not a trace element, but
there is some evidence that phytate decreases its absorption.
If so, one might expect soy protein to decrease Mg availability. Two groups (Erdman et al, 1978 and Lo et al 1978) have
recently examined this question and concluded that soy protein
has no significant effect on Mg availability.

IV.   Constituents of Soybean Protein that may affect
         Bioavailability

Phytate and Zinc. Phytate phosphorus makes up approximately 70% of the total phosphorus in soybeans and it is associated with the protein in soybean products, including the
isolated protein. There is overwhelming evidence that soluble
phytate added to purified diets decreases zinc availability in
chicks (O'Dell and Savage, 1960; Likuski and Forbes, 1964),
pigs (Oberleas et al, 1962), rats (Oberleas et al, 1966;
Likuski and Forbes, 1965) and man (Reinhold et al, 1973). Not
only does phytate bind zinc and make it less readily absorbed
but, as in the case of soybean protein, excess calcium aggra-
vates the situation (Oberleas et al, 1962; O'Dell et al, 1964;
Likuski and Forbes, 1965; Oberleas et al, 1966). In the absence of phytate excess calcium has no effect on zinc availability. It has been postulated that calcium, zinc and phytate interact to form a highly insoluble complex which reduces
the absorption of zinc to a greater extent than phytate alone
(Byrd and Matrone, 1965; Oberleas et al, 1966). The addition
of EDTA to diets containing soluble phytate increases the absorption of zinc (Oberleas et al, 1966; O'Dell et al, 1964)
just as it does when added to soy protein. Data illustrating
the interaction of calcium, zinc and phytate as well as the
counteracting effect of EDTA are shown on Table 5. From these
and similar data, it is clear that absolute values of bioavailability cannot be obtained. Bioavailability of zinc depends, among other things, on the presence of chelating agents
and on the calcium concentration in the diet.

As regards the effect of phytate on zinc availability,
it is important to consider the molar ratio of phytate to
zinc, taking into account the calcium to phytate ratio as
well. From the data presented in Table 5, it appears that 8
ppm of zinc is adequate for the rat if the diet contains no
phytate, but it is grossly inadequate in the presence of 1%
phytate. Such a diet would have a phytate:Zn ratio of 123:1.

Table 5. *Effect of Calcium-Zinc-Phytate Interaction and of EDTA on Zinc Availability in Rats*

Dietary Variables			Weekly Gain, g	
Ca %	Phytate %	EDTA %	Zn Supplement 0	55 ppm
0.8	–	–	32	31
1.6	–	–	33	32
0.8	1.0	–	20	31
1.6	1.0	–	7	33
1.6	1.0	0.1	23	–
1.6	–	0.1	26	–

Data adapted from Oberleas et al (1966). Diet was based on casein and contained approximately 8 ppm Zn. Weanling rats were fed the diets for 4 weeks.

Although there are insufficient data to establish a tolerable ratio, it is probably less than 20:1 (Ellis and Morris, 1978). Zinc phytate itself is well utilized (Green et al, 1962; Ellis and Morris, 1978). In this regard it is significant to point out that high levels of phytate in the diet reduce the biological half-life of zinc in the rat, possibly by inhibiting resorption of endogenous zinc and thus promoting loss of zinc other than that associated with phytate in the diet (Davies and Nightingale, 1975).

Phytate and Iron. There are conflicting results regarding the effect of phytate on iron bioavailability. McCance et al (1943) observed that addition of sodium phytate to white bread decreased iron absorption and they attributed the lower iron balances of persons consuming brown bread to its higher phytate content. In a study with adolescent boys, Sharpe et al (1950) observed that addition of sodium phytate to milk (0.2 g in 200 ml) decreased iron absorption 15-fold. However, the same quantity of phytate in the form of oatmeal had much less effect. In a study with rats, Davies and Nightingale (1975) found that addition of 1% phytate to an egg albumin diet, significantly reduced the whole body retention of iron as well as of zinc, copper and manganese. Others have found little or no adverse effect of dietary phytate on iron utilization (Fuhr and Steenbock, 1943; Cowan et al, 1966; Ranhotra et al, 1974). Morris and Ellis (1976) have found that the iron of monoferric phytate is well utilized, but that insoluble ferric phytate has a much lower bioavailability. These conflicting results suggest that different and undefined conditions existed in the various laboratories. It is possible that phytate interacts with proteins endogenously or when it is added to certain diets. Under conditions of protein-phytate interaction, iron may not be bound efficiently. This concept is supported by the observation of Sharpe et al (1950) that when added to milk the phytate in rolled oats did not prevent iron absorption as effectively as sodium phytate.

Fiber and Zinc. Several studies suggest that phytate is not the only constituent of plant foods that adversely affects zinc utilization. Reinhold et al (1973) observed that unleavened whole wheat (Tanok) bread had a more detrimental effect on zinc balance than sodium phytate supplied daily at the same level as that in bread. However, it should be noted that the phytate was supplied in orange juice separate from the leavened bread. Although this observation could be explained by the chemical form of the phytate, e.g., protein-phytate complexes, there is evidence that fiber decreases absorption of zinc, calcium, magnesium and phosphorus (Reinhold et al, 1976; Ismail-Beigi et al, 1977). In the latter study 10 g of cellulose decreased the retention of these elements when added to a diet that supplied 500 g of bread containing 0.35% of phytate and 3.6% of crude fiber. Addition of cellulose increased the stool size and, no doubt, the rate of passage through the intestinal tract. Whether these changes explain the decreased retention of the elements or there is specific binding to cellulose or other dietary fiber components is not clear. Sandstead et al (1978) tested five sources of dietary fiber, including wheat bran, and concluded that fiber did not alter zinc balance in adult men. Guthrie and Robinson (1978) performed zinc balances in four young women, with and without daily supplements of wheat bran (14 g/day), and found no overall difference in zinc metabolism. Ellis and Morris (1978) found that removal of phytate from wheat bran eliminated its detrimental effect on zinc bioavailability even though the dietary fiber content was unchanged. Caution must be exercised in drawing conclusions relative to the effect of dietary fiber on trace element availability because "dietary fiber" is ill-defined and difficult to determine.

V.   Phytate-Protein Interaction

Phytate forms strong complexes with some proteins and

such complexes are less subject to proteolytic cleavage than the free protein (O'Dell and Boland, 1976). Calcium also interacts with phytate and protein to decrease solubility. In view of these facts it is conceivable that protein-phytate complexes bind zinc and other cations more tenaciously than phytate alone. However, it is clear that protein is not essential for phytate to decrease zinc utilization. Likuski and Forbes (1964) observed that phytic acid decreases the bioavailability of zinc as effectively when amino acids serve as the nitrogen source as when casein is present.

Nevertheless, we have examined extracts of corn germ, sesame meal and defatted soybeans for protein-phytate complexes. Boland et al (1975) observed that the phytate in corn germ is highly water soluble (pH 6.1), that in sesame meal is only slightly water soluble (pH 6.8), while essentially none of that in isolated soy protein is soluble in water (pH 4.9). The distribution of phytate and protein in fractions of defatted soybean flakes is shown in Table 6. Under the conditions used, water extracted approximately 60% of both protein and phytate. Adjustment of the water extract to the isoelectric point (pH 4.5) precipitated slightly more than half of the protein and phytate. Most of the phytate left in the supernate was lost by dialysis.

Table 6. *Protein and Phytate Content of Defatted Soybean Fractions*

Fraction	Solids	Crude Protein	Phytate P Conc.
	g	g	%
Original Flakes	100	49.7	0.43
H$_2$O Sol.	38	29.4 (59%)[1]	0.68 (60%)[1]
Residue	60	15.7 (32%)	0.14 (20%)
Isoelect. Ppt of H$_2$O Sol.	20	18.1 (36%)	0.66 (30%)
Isoelect. Sol. (Dialyzed)	11	1.9 (3.7%)	0.35 (9%)

[1]Percentage of original recovered.

To determine protein-phytate interactions, duplicate
samples of the fractions were electrophoresed on polyacrylamide gels at a running pH of 9.3. One gel was stained for
protein with amidoschwarz and the other reacted with $FeCl_3$
to detect phytate. Corresponding bands indicate protein-
phytate interactions. Figure 3 shows gels of the fractions
prepared from soybean flakes and Figure 4 gives the density
profile of the proteins in the original water extract. Phytate was associated with at least three protein bands and a
large proportion appeared at the buffer front. The major protein band, second from the origin, contained appreciable phytate as did a band which moved near the front. The latter
band appeared in the isoelectric supernate and represents
either low molecular material or protein with a high negative
charge. The isoelectric precipitate contained one band of
phytate which was associated with the major protein near the
origin. Figure 3 also shows similar data for sesame meal.

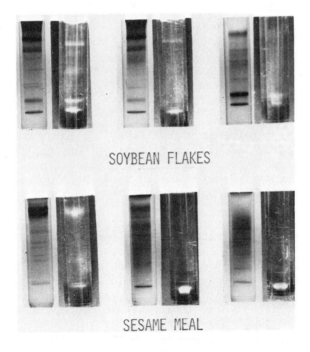

FIGURE 3. Polyacrylamide gel electrophoretograms of defatted soybean extracts. The left one of each pair was
stained with amidoschwarz to detect protein and the right one
was stained with $FeCl_3$ to detect phytate. The first pair on
the left was the total water soluble extract, the center pair,
the isoelectric precipitate and the right hand pair the supernate of the isolectric precipitate.

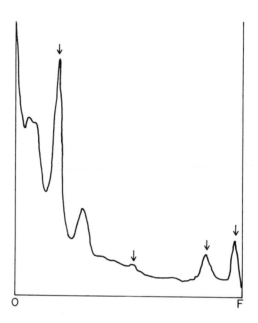

FIGURE 4. The density profile of the water soluble proteins of defatted soybean flakes; gel on the left of Figure 3. 0 corresponds to the origin and F the buffer front of the gel. The arrows indicate protein bands that correspond with phytate bands in the duplicate gel.

The effect of protein-phytate complexes on zinc bio-availability has not been tested. In fact, it is not known whether such complexes exist in seeds or are simply artifacts of preparation. In order for protein-phytate complexes to decrease zinc absorption more than phytate alone, they would have to be indigestible. Against this concept is the observation that the percentage of fecal nitrogen in the chick was not different when soy protein and casein were fed (Savage et al, 1964). However, a relatively small phytate peptide complex might be highly effective in metal binding. While this unresolved problem is of interest, it appears that protein-phytate complexes are of little nutritional significance.

VI.   *Possibilities of Reducing Phytate in Foods*

Soy protein is an excellent source of amino acids for
human and animal nutrition. One limitation to its wide use is
its propensity to bind zinc and decrease zinc absorption. The
phytate associated with soy protein is the major contributor
to zinc binding, and it appears that elimination of this component would improve the value of soy protein as a human
food. This might be accomplished by genetic selection or
chemical treatment. An approach to selection of wheat varieties low in phytate has been made by Bassiri and Nahapetian
(1977) and a similar attempt might be made with soybeans. A
more immediate solution might be achieved by removal of phytate from soy products through the action of phytase, an enzyme that cleaves the phosphate ester bonds of phytate.
Reinhold et al (1974) were able to reduce the phytate content
of bread minimally by use of yeast, but there appears to be an
inhibitor in whole wheat. Soybeans contain little or no endogenous phytase activity. However, Ranhotra et al (1974)
found that at least 75% of the original phytate was hydrolyzed
during the process of making bread from flour enriched with
10% soy flour. Maximal hydrolysis occurred when 9 g of yeast,
the customary amount used, was added per pound loaf. Phytate
can be removed by treatment of the water soluble proteins of
defatted soybeans with an anion exchange resin (Smith and
Rackis, 1957) or by autoclaving (Boland et al, 1975), but neither of these methods appears to be of practical importance.
By selection of the conditions of precipitating soy protein
lower than usual phytate content can be achieved (Ford et al,
1978). Phytate removal was best attained by either low pH
(3.5-4.0) and high calcium (0.04 M) or a higher pH (5.0-5.5)
with low calcium. Using a pH of 5.5 and a calcium concentration of 0.0025 M, Ford et al (1978) prepared a curd from full-
fat soy flour which contained 0.04% phytate phosphorus, only
10% of the original concentration.

CONCLUSIONS

Among the essential trace elements of practical importance
in human nutrition, only zinc has been clearly shown to have
low bioavailability in the presence of soy protein. There is
some evidence that iron, copper and manganese in soy protein
are less well utilized than when consumed in animal proteins.
Soy protein and soy products generally contain 1.0-1.5% phytate, a compound that is clearly implicated in decreasing the
bioavailability of zinc. The detrimental effect of phytate is

aggravated by excess calcium. Under some conditions phytate decreases iron bioavailability, but iron absorption from soy protein generally appears to approach that of animal proteins. Fiber has been implicated in decreasing zinc absorption, but the low concentration of crude fiber in soybean products in general and in soy protein in particular suggests that fiber is responsible for little of the low zinc availability associated with these products. With the proper processing of soy protein so as to lower phytate, the nutritional value of soy protein could be substantially improved.

REFERENCES

Anderson, G. H.; Harris, L.; Rao, A. V. and Jones, J. D. (1976) *J. Nutrition 106*, 1166–1174.

Bassiri, A. and Nahapetian, A. (1977) *J. Agr. Food Chem. 25*, 1118–1122.

Boland, A. R.; Garner, G. B. and O'Dell, B. L. (1975) *Agr. Food Chem. 23*, 1186–1189.

Bowering, J.; Sanchez, A. M. and Irwin, M. I. (1976) *J. Nutrition 106*, 987–1074.

Byrd, C. A. and Matrone, G. (1965) *Proc. Soc. Exp. Biol. Med. 119*, 347–349.

Cartter, J. L. and Hopper, T. H. (1942) *USDA Tech. Bull.* 787.

Cowan, J. W.; Esfahani, M.; Salji, J. P. and Azzam, S. A. (1966) *J. Nutrition 90*, 423–427.

Davies, N. T. and Nightingale (1975) *Brit. J. Nutrition 34*, 243–258.

Davis, P. N.; Norris, L. C. and Kratzer, F. H. (1962) *J. Nutrition 77*, 217–223.

Ellis, R. and Morris, E. R. (1978) *Fed. Proc. 37*, 585 (Abst.).

Erdman, J. W.; Weingartner, K. E.; Parker, H. M. and Forbes, R. M. (1978) *Fed. Proc. 37*, 891 (Abst.).

Evans, G. W. and Johnson, P. E. (1977) *Fed. Proc. 36*, 1100 (Abst.).

Ferretti, R. J. and Levander, O. A. (1976) *J. Agr. Food Chem. 24*, 54–56.

Fitch, C. D.; Harville, W. E.; Dinning, J. S. and Porter, F. S. (1964) *Proc. Soc. Exp. Biol. Med. 116*, 130–133.

Ford, J. R.; Mustakas, G. C. and Schmutz, R. D. (1978) *J. Amer. Oil Chemists Soc. 55*, 371–374.

Forbes, R. M. (1961) *J. Nutrition 74*, 194–200.

Forbes, R. M. (1964) *J. Nutrition 83*, 225–233.

Forbes, R. M. and Parker, H. M. (1977) *Nutrition Reports International 15*, 681-688.

Fritz, J C.; Pla, G. W.; Roberts, T.; Boehne, J. W. and Hove, E. L. (1970) *J. Agri. Food Chem. 18*, 647-651.

Fuhr, I. and Steenbock, H. (1943) *J. Biol. Chem. 147*, 59-64.

Green, J. D.; McCall, J. T.; Speer, V. C. and Hays, V. W. (1962) *J. Animal Science 21*, 997 (Abst.).

Greger, J. L.; Abernathy, R. P. and Bennett, O. A. (1978) *Am. J. Clin. Nutrition 31*, 112-116.

Guthrie, B. E. and Robinson, M. F. (1978) *Fed. Proc. 37*, 254 (Abst.).

Imail-Beigi, F.; Reinhold, J. C.; Faradji, B. and Abadi, P. (1977) *J. Nutrition 107*, 510-518.

Kratzer, F. H.; Allred, J. B.; Davis, P. N.; Marshall, B. J. and Vohra, P. (1959) *J. Nutrition 68*, 313-322.

Lease, J. G.; Barnett, B. D.; Lease, E. J. and Turk, D. E. (1960) *J. Nutriton 72*, 66-70.

Likuski, H. J. A. and Forbes, R. M. (1964) *J. Nutrition 84*, 145-148.

Likuski, H. J. A. and Forbes, R. M. (1965) *J. Nutrition 85*, 230-234.

Lo, G. S. (1978) Personal Communication from Ralston Purina Co.

Lo, G. S.; Collins, D. W.; Steinke, F. H. and Hopkins, D. T. (1978) *Fed. Proc. 37*, 667 (Abst.).

Lolas, G. M. and Markakis, P. (1975) *Agr. Food Chem. 23*, 13-15.

Lolas, G. M.; Palamidis, N. and Markakis, P. (1976) *Cereal Chem. 53*, 867-871.

McCance, R. A.; Edgecombe, C. N. and Widdowson, E. M. (1943) *Lancet 2*, 126-128.

Meyer, E. W. (1978) Personal Communication. Central Soya Co.

Morris, E. R. and Ellis, R. (1976) *J. Nutrition 106*, 753-760.

Morris, E. R. and Ellis, R. (1978) *Proc. 10th National Conference on Wheat Utilization* (In press) also personal communication.

Momcilovic, B.; Belonje, B.; Giroux, A. and Shah, B. G. (1976) *J. Nutrition 106*, 913-917.

Oberleas, D.; Muhrer, M. E. and O'Dell, B. L. (1962) *J. Animal Science 21*, 57-61.

Oberleas, D.; Muhrer, M. E. and O'Dell, B. L. (1966) *J. Nutrition 90*, 56-62.

O'Dell, B. L. (1969) *Amer. J. Clinical Nutrition 22*, 1315-1322.

O'Dell, B. L. (1972) *Annals N. Y. Acad. Sciences 199,* 70-81.

O'Dell, B. L. and Savage, J. E. (1957) *Fed. Proc. 16,* 394 (Abst.)

O'Dell, B. L. and Savage, J. E. (1960) *Proc. Soc. Exp. Biol. Med. 103,* 304-306.

O'Dell, B. L.; Yohe, J. M. and Savage, J. E. (1964) *Poultry Science 43,* 415-419.

O'Dell, B. L.; Burpo, C. E. and Savage, J. E. (1972a) *J. Nutrition 102,* 653-660.

O'Dell, B. L.; Boland, A. R. and Koirtyohann, S. R. (1972b) *J. Agr. Food Chem. 20,* 718-721.

O'Dell, B. L. and Boland, A. R. (1976) *Agr. Food Chemistry 24,* 804-808.

Osborn, T. W. (1977) *J. Agr. Food Chem. 25,* 229-232.

Ranhotra, G. S.; Loewe, R. J. and Pergot, L. V. (1974) *Cereal Chemistry 51,* 323-329.

Ranhotra, G. S.; Loewe, R. J. and Puyat, L. V. (1974) *J. Food Science 29,* 1023-1025.

Reinhold, J. C.; Faradji, B.; Abadi, P. and Ismail-Beigi, F. (1976) *J. Nutrition 106,* 493-503.

Reinhold, J. G.; Nasr, K.; Lahimgarzadeh, A. and Heydayati, H. (1973) *Lancet 1,* 283-288.

Reinhold, J. G.; Parsa, A.; Karimian, N.; Hammick, J. W. and Ismail-Beigi, F. (1974) *J. Nutrition 105,* 976-982.

Sandstead, H.; Klevay, L.; Munoz, J.; Jacob, R.; Logan, G.; Dintzis, F.; Inglett, G. and Shuey, W. (1978) *Fed. Proc. 37,* 254 (Abst.).

Savage, J. E.; Yohe, J. M.; Pickett, E. E. and O'Dell, B. L. (1964) *Poultry Science 43,* 420-426.

Sharpe, L. M.; Peacock, W. C.; Cook, R. and Harris, R. S. (1950) *J. Nutrition 41,* 433-446.

Smith, W. H.; Plumlee, M. P. and Beeson, W. M. (1962) *J. Animal Science 21,* 399-405.

Smith, A. K. and Rackis, J. J. (1957) *J. Amer. Chem. Soc. 79,* 633-637.

Steinke, F. H. and Hopkins, D. T. (1978) *J. Nutrition 108,* 481-489.

Welch, R. M.; House, W. A. and Allaway, W. H. (1974) *J. Nutrition 104,* 733-740.

Welch, R. M. and van Campen, D. R. (1975) *J. Nutrition 105,* 253-256.

DISCUSSION

Erdman: I agree with your overall conclusion that zinc is

the mineral of most concern when considering the availability
of minerals from high phytate foods. At the University of
Illinois in conjunction with Dr. R. M. Forbes we have been
conducting extensive studies concerned with the bioavailability of zinc, iron, calcium and magnesium from soybean foods
using slope ratio assay procedures.

There are several issues that concern phytate and mineral
availability that should be mentioned at this point. First of
all in considering soy protein products, we must be concerned
with the bioavailability of minerals from the soy product itself as well as the effects of the presence of soy protein
upon the bioavailability of minerals from the rest of the
diet. Thus far in our studies we have found that the presence
of soy products in animal diets have little or no effect upon
the bioavailability of inorganic salts added to these diets.
This would suggest that extending chopped meat with soy protein for example should not effect the iron bioavailability
from the meat.

We will present work later this year demonstrating that
the inclusion of soybean hull, which is about 50% fiber, into
soy-flour based diets has no effect upon the bioavailability
of zinc and calcium for the rat.

Finally, I would like to bring up the area of food processing and the formation of insoluble phytic acid - mineral
and phytic acid - mineral - protein complexes. Knowledge of
solubility characteristics of protein, phytic acid and various
phytate salts in various pH regions during processing of soy
protein products may aid in predicting the extent of mineral
binding and, therefore, mineral bioavailability. Perhaps Joe
Rackis would like to speak to this issue, as he has been working in this area for several years.

O'Dell: I wish to emphasize the importance of the phytate
to zinc ratio in considering the effect of phytate on zinc
availability. If the molar ratio becomes greater than 20,
there is likely to be decreased absorption of zinc regardless
of the source of zinc in the diet.

Rackis: I wish to emphasize that reports suggesting soy
protein inhibits the body's ability to utilize zinc has created great controversy. Part of the problem is the failure of
the investigators to make a distinction between flours, concentrates, and isolates and to describe conditions of preparation of the products. Not all problems with mineral bioavailability are associated with phytic acid content *per se*. A
case in point, zinc availability in the isoelectric form of
soy protein isolate, specifically manufactured for use in infant formulas, is high, while some isolates modified by alkali

and other conditions of processing exhibit low bioavailability. Yet, both types have similar amounts of phytic acid. Since such a variety of soy protein products are available, I would suggest that as much information as possible concerning processing history should become a part of all investigations on the nutritional quality of soy products.

O'Dell: I heartily agree that not all problems of mineral bioavailability in foods are related to their phytate content. However, it is clear that the addition to diets of soluble phytate, either the sodium salt or the acid, decreases the availability of zinc. This is the case when it is added to either casein or amino acid based diets. Thus, it is clear that a protein-phytate complex is not essential for zinc binding. It is possible, as we postulated early in our investigations, that some protein-phytate-zinc complexes make zinc less readily absorbed than others. So far as I know, there is no good evidence to support this concept.

Bodwell: For the most part, the few studies such as those of Reinhold which have attempted to clarify the relation between dietary phytate and mineral retention in humans have been confounded by the presence of fiber. A study with primates was done several years ago by Fitch and others in which an undefined source of soy protein was shown to cause severe anemia. Supplementation with iron clearly reversed the effect. This suggests that there may be deleterious effects of dietary phytate on mineral retention in primates including humans. If this is so, it would probably be a serious problem only with the extensive use of meat analogs. This may occur and it is pertinent to suggest that we seriously need some definitive studies with humans to explore the phytate-mineral retention issue. Would you care to comment?

O'Dell: I agree that we need human studies.

Liener: Do you feel that soy protein has a rachitogenic effect?

O'Dell: Dr. Liener's question refers to the observations of Carlson et al, made about 1964, that an isolated soy protein has a rachitogenic effect in turkey poults. The effect was largely overcome by high levels of Vitamin D. Autoclaving the protein did not prevent its rachitogenic effect. It appeared that the protein interfered with absorption of calcium and/or phosphorus. An explanation for the effect is not known, but it is conceivable that a soy protein-phytate complex was involved.

Erdman: The calcium content of soy protein products is quite low.   For example, a rat fed a 20% soy protein diet from full fat soy flour will recieve only 10% of its calcium requirement from the soy.   Soy based diets must be fortified with an additional calcium source.   We have studied the bioavailability of calcium added as calcium carbonate to soy flour, soy beverage and soy concentrate based diets and have found full bioavailability of added calcium.   It is, therefore, safe to assume that rickets development in experimental animals must have been caused by some other factor such as lack of Vitamin D or poorly available phosphorus.

BIOLOGICALLY ACTIVE SUBSTANCES IN SOY PRODUCTS

R. L. Anderson, J. J. Rackis, and W. H. Tallent

"The *toxicity* of a substance is its intrinsic capacity to produce injury when tested by itself. The *hazard* of a substance is its capacity to produce injury under the circumstances of exposure. In connection with the safety of natural food products, our concern is not directly with the intrinsic toxicity of their innumerable chemical components, but rather with the potential hazards of these materials when we eat the foods in which they are present. For example, arsenic, lead, mercury, and fluorine have high intrinsic toxicities, but no hazard is associated with their natural presence in foods. Oxalate is toxic, but its presence in spinach is not a hazard. The cyanogenetic glycoside in lima beans is a highly toxic substance, but it imposes no hazard under the usual conditions of consumption of this food" (Coon, 1973).

As Dr. Coon implies, a wide variety of minor chemical components with diverse biochemical properties occur in the vegetables and plant products we consume as foods. Many of these biologically active substances exhibit toxicity by his definition. Over time, mankind has generally learned to avoid circumstances under which such substances would cause a hazard. We do this through preference for varieties and strains low in the active principles, e.g., the white American variety of lima beans which is at least an order of magnitude lower than other varieties in cyanogenic glycoside content. We also avoid hazards through what Dr. Coon calls "Safety in Numbers." That is, "The wider the variety of food intake, the greater is the number of different chemical substances consumed and the less is the chance that any one chemical will reach a hazardous level in the diet" (Coon, 1973). Finally, there is what he calls "Safety Through Technology." Our grandparents knew that many vegetables had to be soaked, blanched, or cooked in certain ways before they could be eaten without fear of ensuing sickness. Today, modern food processing takes care of much of this concern for us.

Table I. Some Biological Activities of Raw Soybean Meal

Trypsin Inhibition
Hemagglutination
Goitrogenicity
Estrogenicity
Allergenicity
Flatulence

Soybeans are no exception to these generalities. Table I lists the main biological activities of raw soybean meal. Taking trypsin inhibitors (TI's) as an example, careful process control is required to reduce these to low levels without adversely affecting protein quality. Destruction of TI is achieved through treatment with moist heat in an operation called toasting. If conditions are too drastic, chemical reactions involving protein amino acids may be induced. These may lead *inter alia* to decreased nutritional availability of lysine, one of the essential amino acids discussed by Dr. Harper, and to formation (particularly at high pH) of lysinoalanine, the subject of the presentation by Dr. Struthers.
    In the following sections, TI's are discussed in more detail, and the other activities listed in Table 1 are reviewed.

TRYPSIN INHIBITORS

Soy TI's belong to a broad class of proteins which inhibit proteolytic enzymes. Of specific interest is the soybean TI inhibition of trypsin and chymotrypsin, the important animal digestive enzymes for proteins. Several different TI's are present in soybeans, but much of the activity is due to the protein SBTI-A$_2$ (Rackis et al, 1962) which was originally crystallized by Kunitz. The following discussion presents evidence indicating that both the conversion of raw soybeans into products with excellent protein quality and the elimination of the hypertrophic pancreas effect result from the simultaneous destruction of TI's and the transformation of raw protein into a more readily digestible form. Toasting is a very effective common means for destroying soybean TI's. Precise control of heating and other processing variables ensures adequate destruction of these factors to achieve optimum protein quality (American Soybean Assoc., 1974).

*Table II. Summary of Collaborative Study of Trypsin Inhibitor
(TI) Activity of Raw and Toasted Soy Flours[a]*

Sample	Trypsin inhibitor activity		
	TIU[b]/mg sample	mg TI/g sample[c]	% of raw flour
		mg TI/g = $\dfrac{TIU/mg}{1.9}$	
Raw soy flour[d]			
A	92.5	48.7	---
B	105.5	55.5	---
Average	99.0	52.1	100
Toasted soy flour[d,e]			
A	5.5	2.9	6
B	9.4	5.0	10
Average	7.5	4.0	8
Commercial soy flours[f]			
(Toasted)	9.7	5.2	10

[a]*Kakade et al (1974). Average values of several
replicates by four collaborators.*
[b]*TIU = trypsin inhibitor units as defined by Kakade et
al (1969). See text.*
[c]*Calculated on the basis that 1.90 TIU is equivalent to
1 μg of TI, Kakade et al (1969).*
[d]*Prepared in the laboratory from seed-grade soybeans.*
[e]*Live steam at 100°C for 30 minutes.*
[f]*Several lots; heat treatment conditions unknown.*

TI Assay. Conventional assay procedures utilize bovine
trypsin because of its ready availability as a crystalline
preparation; TI content is based on equimolar bovine trypsin
inhibition by soybean TI. The official TI assay procedure of
the American Association of Cereal Chemists and the American
Oil Chemists' Society (Kakade et al, 1974) prescribes soy
sample extraction with 0.01 N sodium hydroxide and dilution of
the extract to a TI concentration sufficient for 40–60% trypsin inhibition. The diluted extract is incubated with bovine
trypsin and a synthetic substrate, benzoyl-DL-arginine-p-ni-
troanilide, for 10 minutes. The absorbance at 410 nm due to

p-nitroaniline is read and converted to trypsin units (TU),
which are arbitrarily defined as an increase of 0.01 absorbance units per 10 ml reaction mixture (Kakade et al, 1969).
TI activity is then expressed in terms of the TU inhibited.

A collaborative study (Kakade et al, 1974) of TI activity
in both raw and toasted defatted soy flours is summarized in
Table II. TI activity can be reported as Trypsin Inhibitor
Units (TIU) per mg sample or protein and as mg TI per g sample
(Table II, footnote c). All trypsin inhibitor activities reported in the present review as TIU per mg of protein were obtained by the method of Kakade et al (1969), those given as
TIU per mg of sample (e.g., flour, concentrate, etc.) by the
improved procedure (Kakade et al, 1974).

Natural Occurrence. Proteinase inhibitors are present
in animal tissues and fluids and in the intestinal tract.
They are widely distributed in the plant kingdom and are present in large amounts in Leguminoseae (peas, beans), Gramineae
(grasses, cereal grains), and Solanaceae (potatoes, egg
plant). Large numbers of species within these families are
important food sources that are eaten either raw or after
cooking. Potatoes, peas, beans, peanuts, soybeans, sweet
corn, wheat, barley, other cereals, and many popular vegetables in the uncooked state have high TI activity (Chen and
Mitchell, 1973).

To determine whether breeding can eliminate TI's and other
antinutritional factors, Kakade et al (1972) analyzed 108 soybean varieties and strains ranging from 66 to 233 TI units per
mg of protein. Raw flours prepared from these soybeans had
low nutritive value and enlarged the rat pancreas. Data for
five of these samples are given in Table III. Toasting was
required in each instance to achieve optimum nutritional quality and to remove the pancreas enlargement effect.

The multiple TI's in soybeans are genetically controlled
variants (Hymowitz, 1973). Investigations by Singh et al
(1969), Hymowitz and Hadley (1972), and Orf and Hymowitz
(1977) on the polymorphic nature of SBTI-A_2 revealed the
presence of three electrophoretically distinguishable forms
designated Ti^1, Ti^2, and Ti^3. These forms are controlled by a codominant multiple allelic system at a single
locus (Hymowitz, 1973). Changes in the three forms of SBTI-
A_2 occur between the 4th and 6th days of germination (Orf et
al, 1977). Frequency of distribution of the TI variants differ in Japanese (Kaizuma and Hymowitz, in press), USDA (Hymowitz, 1973), and European soybean germ plasm (Skorupska and
Hymowitz, in press). A soybean variety that had a TI variant
which was electrophoretically different from the TI in the
usual commercial varieties supported better growth when fed to
rats in the raw state (Yen et al, 1971). Nevertheless, prop-

Table III. Nutritive Value and Trypsin Inhibitor Content of
Different Varieties and Strains of Soybeans[a]

Sample	Protein (%)	Raw flour TIU[b]/mg protein	PER[c,d] Raw	PER[c,d] Toasted[e]
Disoy	39.6	100	1.47	2.66
Provar	41.2	106	1.60	2.20
PI 153319	36.6	168	0.88	2.46
Hark	39.1	100	1.21	2.32
PI 153206	37.5	139	1.21	1.95

[a]Kakade et al (1972).
[b]Trypsin inhibitor units.
[c]Protein efficiency ratio (gram weight gain per gram protein consumed). Data adjusted to a basis of PER = 2.50 for casein.
[d]Pancreas weights were normal in rats fed toasted soy and were negatively correlated with PER in rats fed raw soy.
[e]Autoclaved at 15 psi (120°C) for 30 minutes.

er heat treatment was required to obtain maximum nutritional value for all varieties.

Chicks were used to compare feeding value of raw defatted soybean meal prepared from an experimental strain and from Amsoy 71 variety with commercial toasted defatted meal as a control (Bajjalieh et al, 1977). The strain was found to possess considerably less specific TI activity than Amsoy 71. Raw meal prepared from Amsoy 71 soybeans inhibited growth and enlarged the pancreas to a greater extent (P<0.05) than raw meal from the experimental strain soybeans. Best growth and elimination of pancreatic hypertrophy occurred with commercial toasted soybean meal.

Biological Significance. Affinity chromatography with sepharose bound trypsin was used to remove TI from a raw soy flour extract (Kakade et al, 1973). After comparison of the TI-free and original extracts in rat feeding experiments (Table IV), the authors concluded that approximately 40% of both the growth-depressing and hypertrophic pancreas effects

Table IV. Effect of Trypsin Inhibitor Removal on the Protein
Efficiency Ratio (PER) and Rat Pancreas Weights[a]

Soy extract	TIU/mg protein	PER[b]	Grams pancreas per 100 g body weight[b]
Original	125.1	1.4	0.74
Heated original	13.2	2.7	0.52
Inhibitor-free[c]	12.9	1.9	0.65

[a]*Kakade et al (1973).*
[b]*Data significantly different at P<0.03.*
[c]*Trypsin inhibitor removed by affinity chromatography.*

of the original extract were attributable to the TI. The remaining 60% was due to poor digestibility of the raw protein.
The salient effects of TI, i.e., pancreatic hypertrophy and
growth inhibition, represent an animal's inability to digest
protein and utilize the amino acids in the most efficient manner rather than an irreversible response to a toxic substance;
they can be readily reversed by replacing raw soy flour with
toasted soy flour. After continuous feeding of raw soybean
meal for about one fourth of their lifespan, adult rats maintained body weight and, except for the pancreas, all other organs were normal (Booth et al, 1964). More recently, it was
found that pancreatic hypertrophy in rats fed a diet containing 30% raw soy flour (950 mg TI/100 g diet) can be reversed
by replacing raw flour with toasted flour (99 mg TI/100 g diet). No pancreatic hypertrophy occurred in rats fed food-
grade soy flour, concentrate or isolate for as long as 300
days, about one half of a rat's lifespan (Rackis and Gumbmann,
unpublished data). TI content of the diets ranged from 170 to
320 mg/100 g diet. All organs were normal in size and appearance. A diet of soy milk (plus vitamins and an iron supplement) supported excellent growth, reproduction, and lactation
through at least three generations of rats (Howard et al,
1956).
    Substantial evidence now indicates that the feeding of raw
soybeans and purified soy TI's accelerates protein synthesis
in the pancreas and stimulates hypersecretion of pancreatic
enzymes into the intestinal tract. Continuous stimulation of

the pancreas causes pancreatic hypertrophy and an excessive fecal loss of the protein secreted by the pancreas. In the young animal, growth inhibition occurs. In the adult animal, because of a lower requirement for amino acids, there is no loss of weight, but pancreatic hypertrophy does occur. TI's from lima beans, navy beans, eggs, potatoes, and other foods elicit similar effects (Neiss et al, 1972).

The secretory response of the pancreas to dietary TI is an indirect response that is initiated in the intestine and not in the blood (Olds-Schneeman and Lyman, 1975; Olds-Schneeman et al, 1977). Experiments with rats have demonstrated that pancreatic enzyme secretion is suppressed by negative feedback inhibition resulting from presence of trypsin and chymotrypsin in the intestinal tract (Green and Lyman, 1972; Green et al, 1973; Olds-Schneeman and Lyman, 1975) (Figure 1). The TI's evoke increased pancreatic enzyme secretion by forming inactive trypsin-TI complexes, thereby decreasing the suppression exerted by free trypsin. Feedback inhibition also occurs in humans (Ihse et al, 1977) and pigs (Corring, 1974) but not in dogs (Olds-Schneeman et al, 1977).

A high casein diet (18%) is also a stimulant of pancreatic secretion in the rat, but addition of soy TI to such a diet increases secretion over that of casein alone. Casein forms a complex with trypsin during digestion, thereby decreasing the feedback inhibition and resulting in additional trypsin secretion into the small intestine (Olds-Schneeman et al, 1977). The presence of protein and TI's in the duodenum results in a release of cholecystokinin (CCK) from binding sites in the mucosa. CCK, which has TI activity, is a hormone involved in pancreas regulation. Repeated injections of CCK cause pancreatic hypertrophy and inhibit rat growth. It would appear that TI's and dietary protein stimulate pancreas activity by a common mechanism.

With regard to the question of whether the pancreas is enlarged because of an increase in number of cells (hyperplasia) (Salman et al, 1968; Yanatori and Fujita, 1976; Mainz et al, 1973) or an increase in cell size (hypertrophy) (Booth et al, 1964; Folsch et al, 1974; Booth et al, 1960; Melmed et al, 1973; Beswick et al, 1971), the weight of evidence favors hypertrophy.

It appears that cellular structures are not altered by raw soy flour apart from an increase in acinar cell zymogen granules because of increased protein synthesis. Enzyme secretion from hypertrophied pancreas in direct response to further stimulation by CCK is not blocked or impaired. Soybean TI does not aggravate chemically induced pancreatitis, and pancreatic hypertrophy can be reversed (Booth et al, 1964). All of these facts indicate that TI-induced pancreatic hypertrophy

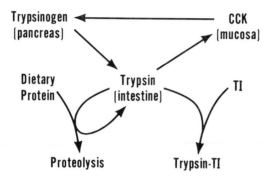

FIGURE 1. Regulation of Trypsin Secretion. CCK, Cholecystokinin; TI, Trypsin Inhibitor.

is a reversible biological response that neither permanently
damages the organ nor impairs its function.

While bovine trypsin is generally used for assays and most
studies have dealt with the effect of TI on animal digestive
processes, human trypsin is only weakly inhibited *in vitro*
by TI (Coan and Travis, 1971; Travis and Roberts, 1969; Feeney
et al, 1969).

Threshold for Nutritional Significance. The relationship between destruction of TI activity and the increase in
PER of defatted soybean meal as a function of toasting is illustrated in Figure 2. No pancreatic hypertrophy occurred in
rats fed soy flour in which only 54% of the TI activity was
destroyed. The diet contained 464 mg TI per 100 g diet (see
diet 4, Table V). The further increase in growth and PER values is attributed to an increase in protein digestibility
rather than to the further destruction of TI activity (Rackis
et al, 1975; Kakade et al, 1973). Maximum PER corresponds to
a destruction of only 80% of the TI activity of raw soy flour
(Rackis et al, 1975). TI content of several commercial soy
protein products is given in Table VI. Residual TI activity
of these products is within the tolerance levels established
in rat bioassay. A number of soy-based infant formulas have
some residual TI activity, and these products likewise do not
cause pancreatic hypertrophy in rats (Churella et al, 1976).

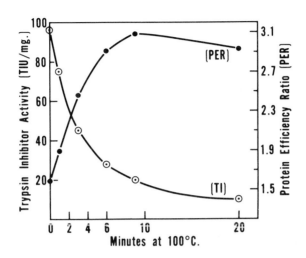

FIGURE 2. Effect of Steaming on Trypsin Inhibitor Activity (TI) and Protein Efficiency Ratio (PER) of Soy Meal. Rackis (1965).

SOYBEAN HEMAGGLUTININ

Phytohemagglutinins, which are named for their characteristic of agglutinating red blood cells, exhibit a specificity toward erythrocytes from various animal species and even toward human red blood cells of various blood groups (Jaffe, 1973). Of 2663 plant species surveyed by Allen and Brilliantine (1969), 800 showed hemagglutinating activity, and many of these hemagglutinins appear to be toxic to animals or to inhibit their growth (Liener, 1974).

Soybeans contain several hemagglutinins (Liener, 1974) comprising an estimated 1-3% of defatted soybean flour (Turner and Liener, 1975; Liener and Rose, 1953). A sevenfold variation in hemagglutinating activity among 108 soybean varieties and strains was reported by Kakade et al (1972). Soy hemagglutinins are glycoproteins of about 110,000 molecular weight. They contain about 5% carbohydrate, consisting mainly of mannose and N-acetyl-D-glucosamine.

Table V. Nutritive Value of Defatted Soy Flour Containing Graded Levels of Trypsin Inhibitor (TI) When Fed to Rats

Diet No.	Dietary protein[a]	TI content mg/100 g diet	Mean body weight (g) ± standard deviation	PER[b]	Nitrogen digestibility[c]	Pancreas weight ± standard deviation g/100 GBW[d]
1	Casein (0)	0	157 ± 16abe	2.50	92	0.48 ± 0.03ce
2	Soy (0)	1001	84 ± 4f	1.13	74	0.68 ± 0.11a
3	Soy (1)	774	94 ± 8ef	1.35	78	0.58 ± 0.01b
4	Soy (3)	464	123 ± 5d	1.75	77	0.51 ± 0.06c
5	Soy (6)	288	141 ± 12c	2.07	83	0.52 ± 0.04c
6	Soy (9)	212	146 ± 11bc	2.19	84	0.48 ± 0.06c
7	Soy (20)	104	139 ± 13c	2.08	83	0.49 ± 0.05c
LSD[f]			11	0.17		0.06

[a] Time (min) of heat treatment at 100°C is given in parentheses.
[b] PER = Protein efficiency ratio corrected on a basis of a PER = 2.50 for casein.
[c] Digestibility = intake–fecal nitrogen/intake X 100.
[d] GBW = grams body weight.
[e] Letters not in common denote statistical significance (P<0.05).
[f] LSD = Least significant difference at the 95% confidence level.

Table VI. *Trypsin Inhibitor (TI) Activity of Various Commercially Manufactured Soy Protein Products*

Product	Trypsin inhibitor activity		Reference
	TIU/mg	% of raw soy flours	
Raw soy flour	99.0	100	b
Toasted soy flour[a]	6-15	6-15	c
Soybean concentrate A	12.0	12	b
Soybean concentrate B	26.5	27	c
Soybean isolate A	8.5	9	b
Soybean isolate B	19.8	20	b
Soybean isolate C	20.9	21	c
Soy food fiber	12.3	12	d
Chicken analog	6.9	7	d
Ham analog	10.2	10	d
Beef analog	6.5	7	d
Textured soy flour	9.8	10	d

[a]*Several lots were analyzed.*
[b]*Kakade et al (1974).*
[c]*Rackis, unpublished data.*
[d]*Liener (1975).*

Soybean hemagglutinins, once thought to be responsible for approximately one half of the rat growth inhibition attributable to raw soybean meal (Liener, 1953), are readily inactivated *in vitro* by heat and by pepsin (Liener, 1958). Inactivation is complete when 12% of the hemagglutinin peptide bonds are cleaved. Birk and Gertler (1961) concluded that hemagglutinin contributed little to the growth impairment of chicks and rats fed raw soybean meal. Two findings led to this conclusion:

a) either HCl (pH 1.6) or pepsin destroyed hemagglutinin, and
b) a raw soybean meal fraction inhibited rat growth only one-
third as much as another which contained just one-seventh the
hemagglutinating activity. The two fractions displayed comparable trypsin inhibitor activities.

Turner and Liener (1975) demonstrated definitively that
soybean hemagglutinin is a relatively minor factor contributing to the poor nutritive value of raw soybeans. Hemagglutinating activity of a crude raw soybean extract was selectively
removed by affinity chromatography on Sepharose-bound conca-
navalin A. The growth rate of young rats was not changed by
hemagglutinin removal but improved, however, with heat treatment of the crude extract. The growth inhibition observed
earlier (Liener, 1953) was evidently due to the *ad libitum*
rat-feeding regimen used, which resulted in a food intake decrease by rats consuming soybean hemagglutinin.

SOYBEAN GOITROGENICITY

It has been apparent for some time that goiters occur in
rats consuming large quantities of soybean meal (McCarrison,
1933; Sharpless et al, 1939; Wilgus et al, 1941). Small
quantities of iodine added to the diet effectively prevent
thyroid enlargement, but the low iodine content of soybeans is
not the only reason for soy goitrogenicity. This was demonstrated by Block and his co-workers (1961) (Table VII) who
found that raw full-fat soy flour produced greater thyroid hypertrophy when fed to rats than did either toasted, defatted
soy flour or isolated soy protein, even though iodine content
was higher in the raw flour than in either the toasted, defatted flour or isolate. Defatting and toasting soy flour diminished the thyroid enlargement effect to that exhibited by soy
protein isolate. Addition of sufficient iodine to supplement
the small quantity of indigenous iodine and to compensate for
the soybean goitrogenic factor effectively prevents goiter
formation in rats fed diets containing soybean. Thyroid enlargement due to soybeans is completely reversible (Block et
al, 1961).

Konijn et al (1972, 1973) have reported a goitrogenic
agent obtained by fractionating soy flour. It appears to be a
peptide comprised of two or three amino acid residues or a
glycopeptide containing one or two amino acid residues linked
to a sugar residue.

Table VII. Goitrogenicity of Soy Products[a]

Product	Iodine content, µg/100 g diet	Thyroid wt. mg/100 g body wt.
Raw soy flour	1.0-2.3	37
Toasted soy flour	0.7	19
Soy isolate	0.9	16
Soy infant food, iodized	40	7
Raw soy flour + 10 µg I_2/g protein	196	8
Casein	30	7

[a]*Block et al (1961).*

The soybean goitrogen mode of action may involve the prevention of intestinal thyroxine readsorption. When raw soy flour was fed to rats with radioactive iodine, an increased thyroid iodine uptake coupled with increased fecal iodine loss was observed (Beck, 1958).

ESTROGENS

The isoflavone glucosides, genistin, daidzin, and glycitein-7-0-β-glucoside (Figure 3) are the major soybean phenolic compounds (Naim et al, 1974). The concentration of each glucoside and its relative estrogenicity are given in Table VIII. Diethylstilbestrol relative estrogenic activity is included for comparison. Friedlander and Sklarz (1971) also detected 6,7,4'-trihydroxyflavone in soybeans, and Knuckles et al (1976) have reported that soybeans contain 120 µg coumestrol/100 g, two thirds of which is extracted with the oil. Although soy isoflavonoids do exhibit estrogenic activity, diethylstilbestrol is 10^5 times more active than genistin

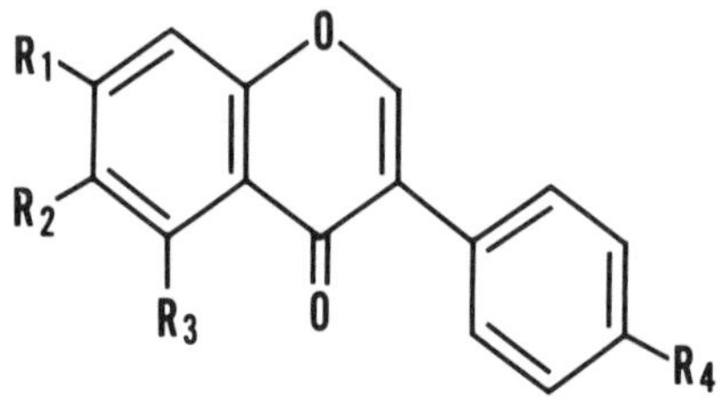

Genistein R_1 = OH; R_2 = H; R_3 = OH; R_4 = OH

Genistin R_1 = O-glucosyl; R_2 = H; R_3 = OH; R_4 = OH

Daidzein R_1 = OH; R_2 = H; R_3 = H; R_4 = OH

Daidzin R_1 = O-glucosyl; R_2 = H; R_3 = H; R_4 = OH

Glycitein R_1 = OH; R_2 = OCH_3; R_3 = H; R_4 = OH

Glycitein-7-β- R_1 = O-glucosyl; R_2 = OCH_3; R_3 = H; R_4 = OH
O-glucoside

FIGURE 3.   Isoflavonoid Compounds in Soybean Meal.

and daidzin is three-fourths as active as genistin (Bickoff et
al, 1962).   Even though coumestrol is 35 times more estrogenic than genistin, coumestrol is much less a factor in soybean
estrogenicity because of its presence in only minute amounts.
The estrogenic activities of the glycitein glucoside and
6,7,4'-trihydroxylflavone are unknown.
     Carter et al (1955) observed little effect on reproduc-
tion in mice fed a control diet containing 0.2% genistin,
while 0.5% produced a definite effect.   In terms of genistin
equivalents (product column, Table VIII), soy meal contains
slightly more estrogenic isoflavone glucosides in total than
were present in the 0.2% genistin diet.   The meal also contains small amounts of the aglycones (genistein and diadzein)
equal to about 1% of the isoflavone glucoside contents.   Depending upon the estrogenicity of glycitein-7-0-β-glucoside,
the estrogenic activity of new soybean varieties could exceed
threshold levels for estrogenic effects.
     Most of the estrogenic activity of soybean meal is attrib-
utable to genistin, which is stable to autoclaving but is
readily extracted by aqueous alcohol.   Soybean meal concen-

Table VIII. *Soy Meal Estrogenicity*

Estrogen	Concentration in soy meal ppm	Relative estrogenicity[a]	Product[b]
Diethylstilbestrol[c]	---	1×10^5	---
Genistin	1644[d]	1.00	1644
Daidzin	581[d]	0.75	436
Glycitein 7-0-β- glucoside	338[d]	---[e]	---
Coumestrol	0.4	35	14

[a]*Bickoff et al (1962).*
[b]*Concentration x estrogenicity.*
[c]*Included for comparison purposes.*
[d]*Naim et al (1974).*
[e]*Estrogenicity unknown.*

trates prepared by alcohol extraction display no estrogenic
activity, and soy protein isolates contain only small amounts
of isoflavones (Nash et al, 1967).

High levels of dietary genistin diminish the growth rates
of mice and rats. Matrone et al (1956) reported that 9 mg
genistin per day per mouse significantly decreased growth rate
to approximately the same extent as 0.64 μg diethylstilbestrol. Greater growth inhibition was observed when large
amounts of genistin were consumed by the mice. Both genistin
and genistein caused significant decreases in weight gain when
fed to rats at the 0.5% level, whereas no effect was noted
when genistin or genistein was included in the diets at the
0.1% level (Magee, 1963). Soybean meal, when used at the same
dietary protein level (19%), would provide only a 0.10% level
of genistin plus genistein and a 0.16% level of total isoflavonoids.

SOYBEAN ALLERGENICITY

As use of soybeans for food becomes more widespread and
consumption of soy proteins increases, problems relating to
soybean sensitivity or allergy might be expected to arise. In
contrast to antinutritional factors and toxins associated with
foodstuffs, allergens display their effects only in those individuals possessing hypersensitivity to allergens; food allergens are innocuous when consumed by most people irrespective of the amount ingested.

There have been several studies to determine whether or
not the kind of protein fed to newborn infants influences the
incidence of allergenic diseases in general (e.g., hay fever,
asthma, chronic rhinitis, etc.) as well as specific food allergies. Glaser and Johnstone (1953) reported that when children from families having allergic diseases were fed soy milk
from birth to 6 months, only 15% of the children developed
some form of allergic disease by 6 years of age, whereas 65%
of the sibling controls and 52% of the nonrelated controls fed
cows' milk developed similar illnesses. In a later study,
Johnstone and Dutton (1966) reported that some form of allergy
occurred in 18% of the soy group and 50% of the control group
fed cows' milk in 235 children from allergic families. On the
other hand, Brown et al (1969) reported allergic responses
from the general population of 11 and 13% for a soy milk-fed
and cows' milk-fed group, respectively. Halpern et al (1973)
monitored 1753 children for various periods to 7 years of age.
These children were fed either breast, soy, or cows' milk from
birth to 6 months. Development of allergy at some time during
childhood was similar in the three milk groups. However, in
the first 6 months 0.5% of the children displayed symptoms of
allergy to soy milk while 1.8% developed allergy to cows' milk.

In attempting to answer the question of why earlier workers reported more allergy in babies fed cows' milk and found
soy milk to be relatively hypoallergenic, Halpern et al (1973)
suggest that with better control of processing, including
higher and longer heat treatment, cows' milk is now considerably less allergenic.

Five newborn infants developed eczema on first postnatal
contact with soybeans, apparently as a result of intrauterine
sensitization (Kuroume et al, 1976). Apparently both prena-
tal and postnatal sensitization to soybeans are possible.
Whitington and Gibson (1977) described soy protein intolerance
after oral challenge with soy protein isolate in four infants. Responses to the challenge included diarrhea, vomiting, hypertension, lethargy, and fever. Each of the infants

was also sensitive to cows' milk protein. Four children were found by Halpin et al (1977) to display soy formula intolerance. The children experienced prolonged diarrhea, weight loss, vomiting, and fever. They had been receiving soy formula for suspected cows' milk allergy. The authors concluded that soy formulas are not hypoallergenic. Nevertheless, soybeans appear to be hypoallergenic in the sense that less of the population evidently displays sensitivity to soy than to either other legumes or to cows' milk. On the other hand, some individuals are found who react to soy products while cows' milk causes them no difficulty.

A weak soybean allergen has been isolated in crude form (Spies et al, 1951). To completely destroy the allergenic activity of raw soy meal extracts requires heating for 30 min. at 180°C (Perlman, 1965). This is more heat than that required to inactivate the antinutritional factors in raw soy flours and may be sufficient to lower nutritional quality.

FLATULENCE

Food legume flatulence problems have been reviewed recently (Rackis, 1975, 1976; Cristofaro et al, 1974). Due to the absence of α-galactosidase in the human small intestine, galactosido-oligosaccharides such as raffinose, stachyose, and verbascose are not digested. Fermentation of these sugars by intestinal microorganisms in the large intestine then produces large amounts of CO_2 and H_2. Any carbohydrate which, for any reason, escapes digestion and absorption in the small intestine and ferments in the large intestine can contribute to flatus.

Flatus activity of soy protein products in man is shown in Table IX. Both full-fat and defatted soy flours cause increased flatulence, whereas protein isolate and high-molecular-weight polysaccharides (water-insoluble residue) show no more flatus activity than the basal diet. When soy flour is extracted with 80% ethanol to produce a protein concentrate, the flatulence effects are reduced. Alcohol extractives and soy whey solids, which contain most of the oligosaccharides, are responsible for large amounts of flatus.

Smiley et al (1976) describe an immobilized enzyme technique involving use of a hollow-fiber reactor (Figure 4) for removing milk oligosaccharides. Soy milk is circulated around hollow fibers which contain α-galactosidase. Raffinose and stachyose diffuse into the fibers and are hydrolyzed, thereby

Table IX. Effects of Soy Products on Flatus in Man[a]

Product[b]	Daily Intake (g)	Flatus volume (cm^3/hr) Average	Range
Full-fat soy flour	146	30	0-75
Defatted soy flour	146	71	0-290
Soy protein concentrate	146	36	0-98
Soy proteinate	146	2	0-20
Water-insoluble residue[c]	146	13	0-30
Whey solids[d]	48	300[e]	---
80% Ethanol extractives[d]	27	240	220-260
Basal diet	146	13	0-28

[a]*Data from Steggerda et al (1966), except for whey solids and 80% ethanol extractives from Rackis et al (1970).*
[b]*All products were toasted with live steam at 100°C for 40 min.*
[c]*Fed at a level three times higher than that present in the defatted soy flour diet.*
[d]*Amount equal to that present in 146 g of defatted soy flour.*
[e]*One subject, otherwise four subjects per test.*

freeing the soy milk of these oligosaccharides. The partial purification and characterization of the enzyme preparation employed was reported by McGhee et al (1978). It is produced by *Aspergillus awamori* and contains both α-galactosidase and invertase activity.

Raffinose and stachyose contents of soybean varieties differ. The elimination of these oligosaccharides by plant breeding (Hymowitz and Collins, 1974) is a possibility. However, incorporation of such genetic changes into new productive crop varieties usually requires well over a decade, so this must be viewed as a long-term approach at best.

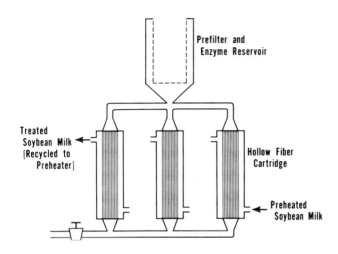

FIGURE 4.   Diagrammatic Sketch of Hollow-Fiber Reactor.
Soybean Milk Supply Line was Connected in Parallel to Fiber
Cartridges.   The Recycle Flow Rate was 185 ml/min., and the
Temperature was 45°C.

The amount of oligosaccharides that can be removed from
mature whole soybeans by leaching is summarized in Table X.
Germination also reduces soy oligosaccharide content (Table
XI).   Calloway et al (1971) observed no change in human flatus activity on germinating various beans, including soy.
Becker et al (1974) reported that the disappearance of raffinose and stachyose during autolysis of California white beans
reduced flatulence, as measured by hydrogen production in the
rat, and that other components may also cause flatulence.
With respect to soybeans, these results would appear to indicate that the high-molecular-weight polysaccharides, which
normally do not cause flatulence, may have been partially hydrolyzed during germination.   These modified polysaccharides
may now become substrates for the formation of flatus by the
intestinal microflora and, thereby, compensate for the loss in
raffinose and stachyose.   On the other hand, the failure to
eliminate flatus activity in the mung beans and other food legumes during germination may be related to the presence of un-
digested starch granules that enter the lower intestinal
tract.   Immature soybeans contain numerous starch granules
which disappear during final days of maturity (Bils and

Table X. *Protein Loss and Oligosaccharide Removal From*
Whole Soybeans by Various Treatments

Treatment	Bean to water ratio	Protein loss (%)[a]	Oligosac-charide removal (%)[b]	Reference
Soak, rm temp, 15 hr	---		8	c
Boil, 20 min, water	1:10	1.0	33	d
Boil, 60 min, water	1:10	2.6	59	d
Boil, 20 min, 0.5% NaHCO$_3$	1:10	1.3	21	d
Boil, 60 min, 0.5% NaHCO$_3$	1:10	6.8	60	d
Boil, 60 min, pH 4.3	1:10	2.0	46	d

[a]*Grams protein/100 g protein in original dry bean.*
[b]*Grams oligosaccharide/100 g oligosaccharide in original*
dry bean.
[c]*Kim et al (1973).*
[d]*Ku et al (1976).*

Howell, 1963). Whether immature soybeans caused flatulence is
not known.

PROTEIN FOR SEVEN BILLION

     What is the overall impact of the biological activities
listed in Table I on growing food use of soy proteins?
Hemagglutination is of no nutritional significance because

Table XI. *Effect of Germination on Autolysis of Sucrose, Raffinose, and Stachyose*

Germination period (hr.)[a]	% Loss of oligosaccharides			Reference
	Sucrose	Raffinose	Stachyose	
48	0	30	50	b
96	80	100	96	b
96	--	80	80	c
120	--	100	90	c

[a]*After imbibition.*
[b]*East et al (1972).*
[c]*Adjei-Twum et al (1976).*

soybean hemagglutinins are rapidly deactivated by heat treatment during processing and by gastric enzymes and acidity. Goitrogenicity is effectively counteracted by sufficient iodine in the diet. Estrogenicity is of such a low level as to be a possible problem only in the case of livestock and only under the rare circumstance of a ration comprised almost completely of soybeans or soybean meal. Like any other food product, increasing consumption of soy proteins will result in more cases of allergenicity, but the weight of total evidence to date still supports their reputation of relative hypoallergenicity. Flatulence causes discomfort and embarrassment but is not antinutritional. We devoted the most attention in this review to trypsin inhibition because of its physiological complexity. Yet, taken in the context that many of our foods contain TI's and that proper processing reduces soybean TI to a level representing insignificant addition to our background intake of TI's from all dietary sources, translating this complexity into cause for serious concern seems unwarranted. The perspective outlined in the opening two paragraphs of this review, with the help of quotations from Coon (1973), is very relevant to the soybean TI issue.

While projections for market growth of soy protein products vary, there seems to be little doubt about the potential of the golden bean to play an important role in meeting the nutritional needs of a growing world population. Just how

much of this potential to help feed seven billion people by
the year 2000 A.D. will be realized depends on our ability to
maintain and improve the quality and economy of soy-based
foods, as well as on consumer confidence in and acceptance of
them. Diligent attention to the chemical, physical, and biological properties of the principles discussed in this paper
will help ensure that these objectives are achieved as diverse
new products and more efficient processes are developed in the
coming decades.

REFERENCES

Adjei-Twum, D. C.; Splittstoessor, W. E.; and Vandemack,
J. S. 1976. Hort. Sci. 11, 235-236.

Allen, N. K., and Brilliantine, L. 1969. J. Immunol.
102, 1295-1299.

American Soybean Association. 1974. Proc. World Soy Protein Conf., Munich, Germany, 1973. J. Am. Oil Chem. Soc. 51,
No. 1.

Bajjalieh, N. L.; Orf, J. H.; Hymowitz, T.; and Jensen, A.
H. 1977. 69th Annual Meeting Am. Soc. Anim. Sci., Abstract
No. 198.

Beck, R. N. 1958. Endocrinology 62, 587-592.

Becker, R.; Olson, A. C.; Frederick, D. P.; Kon, S.; Gumbmann, M. R.; and Wagner, J. R. 1974. J. Food Sci. 39, 766-
769.

Beswick, I. P.; Pirola, R. C.; and Bouchier, I. A. D.
1971. Br. J. Exp. Pathol. 52, 252-255.

Bickoff, E. M.; Livingston, A. L.; Hendrickson, A. P.; and
Booth, A. N. 1962. J. Agric. Food Chem. 10, 410-412.

Bils, R. F. and Howell, R. W. 1963. Crop Sci. 3, 304-
308.

Birk, Y. and Gertler, A. 1961. J. Nutr. 75, 379-387.

Block, R. J.; Mandl, R. H.; Howard, H. W.; Bauer, C. D.;
and Anderson, D. W. 1961. Arch. Biochem. Biophys. 93, 15-24.

Booth A. N.; Robbins, D. J.; Ribelin, W. E.; and DeEds, F.
1960. Proc. Soc. Exp. Biol. Med. 104, 681-683.

Booth, A. N.; Robbins, D. J.; Ribelin, W. E.; DeEds, F.;
Smith, A. K.; and Rackis, J. J. 1964. Proc. Exp. Biol. Med.
116, 1067-1069.

Brown, E. B.; Josephson, B. M.; Levine, H. S.; and Rosen,
M. 1969. Am. J. Dis. Child. 117, 693-698.

Calloway, D. H.; Hickey, C. A.; and Murphy, E. L. 1971.
J. Food Sci. 36, 251-255.

Carter, M. W.; Matrone, G.; and Smart, W. W. G., Jr.
1955. J. Nutr. 55, 639-645.

Chen, I. and Mitchell, H. L. 1973. Phytochemistry 12, 327-330.

Churella, H. B.; Yao, B. C.; and Thomson, W. A. B. 1976. J. Agric. Food Chem. 24, 393-396.

Coan, M. H. and Travis, J. 1971. "Proc. Int. Res. Conf. Proteinase Inhibitors," (H. Fritz and H. Tschesche, eds.), pp. 294-298. W. deGruyter, Berlin.

Coon, J. M. 1973. "Toxicants Occurring Naturally in Foods," 2nd ed., (Committee on Food Protection, Food and Nutrition Board), pp. 573-594. National Research Council, National Academy of Sciences, Washington, DC.

Corring, T. 1974. An. Biol. Biochem. Biophys. 14, 487-498.

Cristofaro, E.; Mottu, F.; and Wuhrmann, J. J. 1974. "Sugars in Nutrition," (H. L. Sipple and K. W. McNutt, eds.), pp. 313-336. Academic Press, New York.

East, J. W.; Nakayama, T. O. M.; and Parkman, S. B. 1972. Crop Sci. 12, 7-9.

Feeney, R. E.; Means, G. E.; and Bigler, J. C. 1969. J. Biol. Chem. 244, 1957-1960.

Folsch, U. R.; Winckler, K.; and Wormsley, K. A. 1974. Digestion 11, 161-171.

Friedlander, A. and Sklarz, B. 1971. Experientia 27, 762-763.

Glaser, J. and Johnstone, D. E. 1953. J. Am. Med. Assoc. 153, 620-622.

Green, G. M.; Olds, B. A.; Matthews, G.; and Lyman, R. L. 1973. Proc. Soc. Exp. Biol. Med. 142, 1162-1167.

Green, G. M. and Lyman, R. L. 1972. Proc. Soc. Exp. Biol. Med. 140, 6-12.

Halpern, S. R.; Sellars, W. A.; Johnson, R. B.; Anderson, D. W.; Saperstein, S.; and Reisch, J. S. 1973. J. Allergy Clin. Immunol. 51, 139-151.

Halpin, T. C.; Byrne, W. J.; and Ament, M. E. 1977. J. Pediatr. 91, 404-407.

Howard, H. W.; Block, R. J.; Anderson, D. W.; and Bauer, C. D. 1956. Ann. Allergy 14, 166-169.

Hymowitz, T. 1973. Crop Sci. 13, 420-421.

Hymowitz, T. and Collins, F. I. 1974. Agron. J. 66, 239-241.

Hymowitz, T. and Hadley, H. H. 1972. Crop Sci. 12, 197-198.

Ihse, I.; Lilja, P.; and Lundquist, I. 1977. Digestion 15, 303-308.

Jaffee, W. G. 1973. "Toxicants Occurring Naturally in Foods." 2nd ed., (Committee on Food Protection, Food and Nutrition Board), pp. 106-129. National Research Council, National Academy of Sciences, Washington, DC.

Johnstone, D. E. and Dutton, A. M. 1966. New Engl. J. Med. 274, 715-719.

Kaizuma, N. and Hymowitz, T. In press. Jpn. J. Breed.

Kakade, M. L.; Simons, N. R.; and Liener, I. E. 1969. Cereal Chem. 46, 518-526.

Kakade, M. L.; Simons, N. R.; Liener, I. E.; and Lambert, J. W. 1972. J. Agric. Food Chem. 20, 87-90.

Kakade, M. L.; Hoffa, D. E.; and Liener, I. E. 1973. J. Nutr. 103, 1772-1778.

Kakade, M. L.; Rackis, J. J.; McGhee, J. E.; and Puski, G. 1974. Cereal Chem. 51, 376-382.

Kim, W. J.; Smit, C. J. B.; and Nakayama, T. O. M. 1973. Lebensm.-Wiss. Technol. 6, 201-204.

Konijn, A. M.; Edelstein, S.; and Guggenheim, K. 1972. J. Sci. Food Agric. 23, 549-555.

Konijn, A. M.; Gershon, B.; and Guggenheim, K. 1973. J. Nutr. 103, 378-383.

Knuckles, B. E.; deFremery, D.; and Kohler, G. O. 1976. J. Agric. Food Chem. 24, 1177-1180.

Ku, S.; Wei, L. S.; Steinberg, M. P.; Nelson, A. I.; and Hymowitz, T. 1976. J. Food Sci. 41, 361-364.

Kuroume, T.; Oguri, M.; Matsumura, T.; Swasaki, I.; Kanbe, Y.; Yamada, T.; Kawabe, S.; and Negishi, K. 1976. Ann. Allergy 37, 41-46.

Liener, I. E. 1953. J. Nutri. 49, 527-539.

Liener, I. E. 1958. J. Biol. Chem. 233, 401-405.

Liener, I. E. 1974. J. Agric. Food Chem. 22, 17-22.

Liener, I. E. 1975. "Protein Nutritional Quality of Foods and Feeds," (M. Friedman, ed.), pp. 523-550, Vol. 1, Part 2, Dekker, Inc., New York.

Liener, I. E. and Rose, J. E. 1953. Proc. Soc. Exp. Biol. Med. 83, 539-544.

Magee, A. C. 1963. J. Nutr. 80, 151-156.

Mainz, D. L.; Black, O.; and Webster, P. D. 1973. J. Clin. Invest. 52, 2300-2304.

Matrone, G.; Smart, W. W. G., Jr.; Carter, M. W.; Smart, V. W.; and Garren, H. W. 1956. J. Nutr. 59, 235-241.

McCarrison, R. 1933. Indian J. Med. Res. 21, 179-181.

McGhee, J. E.; Silman, R.; and Bagley, E. B. 1978. J. Am. Oil Chem. Soc. 55, 244-247.

Melmed, R. N.; Turner, R. C.; and Holt, S. J. 1973. J. Cell Sci. 13, 279-295.

Naim, M.; Gestetner, B.; Zilkah, S.; Birk, Y.; and Bondi, A. 1964. J. Agric. Food Chem. 22, 806-810.

Nash, A. M.; Eldridge, A. C.; and Wolf, W. J. 1967. J. Agric. Food Chem. 15, 102-108.

Neiss, E.; Ivey, C. A.; and Nesheim, M. C. 1972. Proc. Soc. Exp. Biol. Med. 140, 291-296.

Olds-Schneeman, B.; Chang, L.; Smith, L. B.; and Lyman, R. L. 1977. J. Nutr. 107, 281-288.

Olds-Schneeman, B. and Lyman, R. L. 1975. Proc. Soc. Exp. Biol. Med. 148, 897-903.

Orf, J. H. and Hymowitz, T. 1977. Crop Sci. 17, 811-813.

Orf, J. H.; Mies, D. W.; and Hymowitz, T. 1977. Bot. Gaz. 138, 255-260.

Perlman, F. 1965. Food Technol. 20, 1438-1445.

Rackis, J. J. 1965. Fed. Proc. 24, 1488-1493.

Rackis, J. J. 1975. "Physiological Effects of Food Carbohydrates," (A. Jeanes and J. Hodge, eds.), pp. 207-222, American Chemical Society, Washington, DC.

Rackis, J. J. 1976. "World Soybean Research," (L. D. Hill, ed.), pp. 892-903, The Interstate Printers and Publishers, Inc., Danville, Illinois.

Rackis, J. J.; McGhee, J. E.; and Booth, A. N. 1975. Cereal Chem. 52, 85-92.

Rackis, J. J.; Sasame, H. A.; Mann, R. K.; Anderson, R. L.; and Smith, A. K. 1962. Arch. Biochem. Biophys. 98, 471-478.

Rackis, J. J.; Sessa, D. J.; Steggerda, F. R.; Shimizu, J.; Anderson, J; and Pearl, S. L. 1970. J. Food Sci. 35, 634-639.

Salman, A. J.; Pubols, M. H.; and McGinnis, J. 1968. Proc. Soc. Exp. Biol. Med. 128, 258-261.

Sharpless, G. R.; Pearsons, J.; and Prato, G. S. 1939. J. Nutr. 17, 545-555.

Singh, L.; Wilson, C. M.; and Hadley, H. H. 1969. Crop Sci. 9, 489-491.

Skorupska, H. and Hymowitz, T. In press. Genet. Pol.

Smiley, K. L.; Hensley, D. E.; and Gasdorf, H. J. 1976. Appl. Environ. Microbiol. 31, 615-617.

Spies, J. R.; Coulson, E. J.; Chambers, D. C.; Bernton, H. S.; Stevens, H.; and Shimp, J. H. 1951. J. Am. Chem. Soc. 73, 3995-4001.

Steggerda, F. R.; Richards, E. A.; and Rackis, J. J. 1966. Proc. Soc. Exp. Biol. Med. 121, 1235-1239.

Travis, J. and Roberts, R. C. 1969. Biochemistry 8, 2884-2889.

Turner, R. H. and Liener, I. E. 1975. J. Agric. Food Chem. 23, 484-487.

Whitington, P. F. and Gibson, R. 1977. Pediatrics 59, 730-732.

Wilgus, H. S., Jr.; Gassner, F. X.; Patton, A. R.; and Gustavson, R. G. 1941. J. Nutr. 22, 43-52.

Yanatori, Y. and Fujita, T. 1976. Arch. Histol. Jpn. 39, 67-78.

Yen, J. T.; Hymowitz, T.; and Jensen, A. H. 1971. J. Anim. Sci. 33, 1012-1017.

LYSINOALANINE: BIOLOGICAL EFFECTS AND SIGNIFICANCE

Barbara J. Struthers, Robert R. Dahlgren, Daniel T. Hopkins
and Marvin L. Raymond

Lysinoalanine (LAL), an unusual amino acid formed as a result of alkali treatment of protein, was first reported in
alkali-treated ribonuclease (Patchornik and Sokolovsky, 1964).
The proposed mechanism of formation is shown in Figure 1
(Feeney, 1977). Dehydration of either serine or cysteine to
dehydroalanine, an α, β unsaturated amino acid, is followed
by addition of the epsilon amino group of lysine to the double
bond. The dehydroalanine precursor is found in some natural
biological systems, for example, as a catalytically essential
amino acid in yeast phenylalanine ammonia lyase (Hodgins,
1971) and as a constituent of the antibiotics cinnamycin and
duramycin which is necessary for antibiotic activity (Gross et
al, 1973). Other crosslinked amino acids besides LAL are
known to form during alkali treatment of protein: ornithino-
alanine (Ziegler et al, 1967) and lanthionine (Horne et al,
1941). However, LAL has received the most attention because
of its ability to induce a unique renal lesion in rats.

CHARACTERIZATION OF NEPHROCYTOMEGALY

The renal lesion was initially observed by Newberne and
Young (1966). Woodard (Woodard and Alvarez, 1967; Woodard,
1969) was the first investigator to fully describe the lesion,
which he termed cytomegalia. Nephrocytomegalia is characterized by enlarged cells, and enlarged and/or multiple nuclei
and nucleoli occurring in proximal tubules in the *pars recta,*
or medullary outer stripe, region of the kidney (Figure 2).
Woodard and Alvarez (1967) correctly associated feeding of alkali treated non-food grade protein with nephrocytomegaly and
later identified LAL formed by alkaline treatment of protein
as the causative agent (Woodard and Short, 1973; Reyniers et

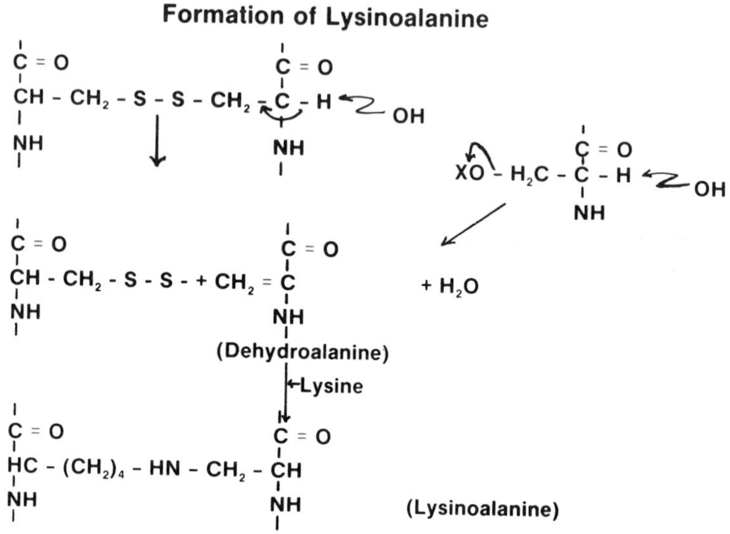

Formation of Lysinoalanine

FIGURE 1.   Formation of lysinoalanine from cysteine or serine and lysine (after Feeney, 1977).

al, 1974).   (Proteins which undergo sufficiently severe alkali treatment to produce nephrocytomegaly are usually of an industrial grade, used for paper coatings, adhesives, etc., and are not intended for human consumption. However, severe alkali treatment of any protein can produce considerable LAL.)   Histochemical investigations by Woodard's group showed an increase in both DNA and associated histones when LAL was fed, with the acid dye binding ratio of arginine and lysine being significantly altered in the megalic nuclei (Reyniers et al, 1974). No loss of kidney function has ever been seen except when extremely high dietary levels (10,000 ppm) of LAL are fed as the free amino acid (DeGroot et al, 1975).   Woodard found the intraperitoneal injection of 30 mg (approximately equivalent to 3,000 ppm) LAL daily for one week caused loss of the

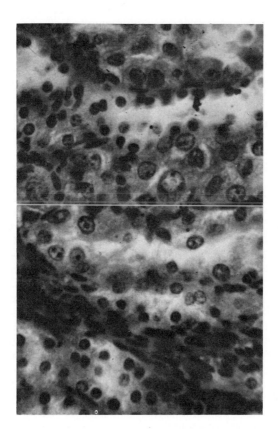

FIGURE 2. Nephrocytomegalic cells from *pars recta* of kidney of rat fed 3,000 ppm protein-bound LAL for eight weeks (Struthers, et al, 1977).

tubule brush border in the affected areas. Nephrocalcinosis, a condition in which calcareous deposits are found in kidneys, has also been associated with feeding of alkali-treated proteins; since this condition frequently occurs independently of nephrocytomegalia (Woodard, 1971), it appears that this is not directly related to LAL (Woodard, 1971; Kaunitz and Johnson,

1976).

The initial work on renal cytomegaly and karyomegaly (nuclear enlargement) was done without knowledge of the amount of
LAL in the alkali-treated protein. The first quantitative
work was done by Woodard (1975, 1977; Woodard et al, 1975),
who fed 250-3,000 ppm free synthetic LAL to weanling Sprague-
Dawley rats and obtained nephrocytomegaly at all levels, with
the higher doses producing the more severe lesions. Intraperitoneal injection of free LAL, 30 mg/day for seven days
(equivalent to 2,000-2,500 ppm free dietary LAL), resulted in
lesions quantitatively and qualitatively similar to those obtained in feeding studies, demonstrating that the free amino
acid is well absorbed (Woodard et al, 1975, 1977).

COMPARISONS OF FREE AND BOUND LAL

Although Woodard consistently found nephrocytomegaly after
feeding alkali-treated protein to Sprague-Dawley derived rats,
DeGroot and Slump (1969) fed dietary LAL levels of 680 and
1,190 ppm from alkali-treated protein to Wistar rats and were
unable to observe any renal cell cytomegaly. When they fed
free synthetic LAL, dietary levels as low as 100 ppm resulted
in production of nephrocytomegaly, demonstrating that there
was considerable difference in the apparent bioavailability of
free and protein-bound LAL. The only difference in the two
groups' experiments with intact protein appeared to be rat
strain. We (Struthers et al, 1977) fed alkali-treated soy
protein diets containing graded levels of LAL from 500-3,000
ppm to Wistar and Sprague-Dawley rats and observed cytomegaly
only at the 3,000 ppm level. The severity of nephrocytomegalia was much greater in the Sprague-Dawley rats; on a scale of
0-3 (normal-severe) the Wistar rats scored 0.6 and Sprague-
Dawley rats, 1.8. A strain difference was also noted in the
extent of nephrocalcinosis developed. On a 0-3 scale, Wistar
rats scored twice as high as Sprague-Dawley rats. We have
noted not only a strain, but a substrain difference in susceptibility to nephrocytomegaly. We consistently find 100% incidence of nephrocytomegalia after eight weeks on our 3,000 ppm
protein-bound LAL diet in Sprague-Dawley derived rats from
Charles River Laboratories[1], and zero incidence in Sprague-
Dawley derived rats from Hilltop Laboratory Animal Company[2].

[1]Charles River Breeding Laboratories, Wilmington, MA.
[2]Hilltop Laboratory Animal Company, Scottsdale, PA.

Table 1. Rat Source and Susceptibility to Nephrocytomegaly

Rat Source	Incidence of Nephrocytomegaly	
	RP-100[1]	ATSP[2]
Hilltop	0/5	0/5
Charles River *CD*	0/5	5/5
Harlan	0/5	2/5
Purina Farm	0/5	3/5

[1]*RP-100: soybean assay protein, Ralston Purina Company.*
[2]*ATSP: alkali-treated soy protein diet (3,000 ppm LAL).*

Sprague-Dawley derived rats from our Research Farm[3] and from Harlan Industries[4] show intermediate incidence levels (Table 1) (Struthers and Dahlgren, unpublished observations). A summary of studies in which free or protein-bound LAL has been fed is shown in Table 2.

There is considerable difference in minimum effect level of free and bound LAL. The greater nephrocytomegalic potency of free LAL compared to protein-bound LAL is evident not only from results of feeding synthetic LAL but from protein hydrolysis studies conducted by DeGroot et al (1976). Feeding of alkali-treated casein containing 2,000 or 6,000 ppm LAL caused no alteration of renal histology. When the acid hydrolysate of these caseins was fed at a level which provided only 200 ppm free LAL, it did produce nephrocytomegaly and the blood LAL level was comparable to that obtained by feeding of 200 ppm synthetic LAL.

The great differences in effect level of feeding free and bound LAL are reflected in the excretion pattern where this has been measured. Table 3 shows that only 1% of excreted LAL fed in protein-bound form is excreted in urine; 99% is fecal (DeGroot et al, 1976). When the same quantity of LAL is fed as the free compound, more than three times as much LAL is re-

[3]Ralston Purina Research Farm, Gray Summit, MO.
[4]Harlan Industries, Indianapolis, IN.

Table 2. Summary of nephrocytomegaly induced by feeding free or bound LAL to rats.

LAL Source	Ppm Dietary LAL Free	Bound	Rat Strain[1]	Duration of Experiment, Weeks	Nephro-cytomegaly[2]	Reference
Synthetic	10	-	W	13	-	DeGroot et al (1976)
	30	-	W	13	-	"
	100	-	W	13	+	"
	200	-	W	4	+	"
	250	-	SD-C	4	+	Woodard (1975)
	500	-	SD-C	4	+	"
	1,000	-	SD-C	4	+	"
	1,000	-	W	4	+	DeGroot (1976)
	2,000	-	W	4	+	" (1976, 1977)
	3,000	-	W	4	++	"
	3,000	-	SD-C	4	++	Woodard (1975)
	10,000	-	W	4	++	DeGroot et al (1976)
AT-casein[3] hydrol.	200	-	W	4	+	"
AT-casein[3]	-	6,000	W	4	+	"
AT-lactalbumin[3]	-	5,000	SD	4	+	Gould & MacGregor (1977)

Table 2. (Cont.)

LAL Source	Ppm Dietary LAL Free	Bound	Rat Strain[1]	Duration of Experiment, Weeks	Nephro-cytomegaly[2]	Reference
AT-soy protein[3] isol.	–	3,000	W	8	+(0.6)	Struthers et al (1977)
	–	3,000	SD-C	8	++(1.8)	Struthers et al (1977)
	–	3,000	SD-H	8	–	"
AT-lactalbumin[3]	–	2,500	SD	4	–	Gould & MacGregor (1977)
	–	2,000	SD-C	8	–	Struthers et al (1977)
	–	2,000	SD-H	8	–	" (1977)
AT-[3]Casein	–	2,000	W	4	–	DeGroot (1976)
AT-[3]Soy prot. isol.	–	1,000	SD-C	8	–	Struthers et al (1977)
	–	1,000	SD-H	8	–	" (1977)
		500	SD-C	8	–	" (1977)
	–	500	SD-H	8	–	" (1977)
AT-Casein[3]	15	200	W	4	–	DeGroot (1976)
	15	–	W	4	–	"
	30	400	W	4	+	"
	45	–	W	4	–	"
	30	–	W	4	–	"

[1]W: Wistar; SD-C: Sprague-Dawley, Charles River; SD-H: Sprague Dawley, Hilltop Animal Laboratories.
[2]+: Present; ++: Present, severe; -: Absent.
[3]AT: alkali-treated.

Table 3. Urinary and fecal excretion of LAL in rats fed intact or hydrolyzed alkali-treated soy protein (ATSP).[1]

LAL Source[2]	Percent of Ingested LAL Recovered In:	
	Urine	Feces
Intact ATSP	0.2	50
Acid-hydrolyzed ATSP	25	7

[1]*DeGroot et al, 1976.*
[2]*Diets contained 2,400 ppm LAL.*

covered in urine as in feces. Finot (1977) measured fecal LAL after feeding several types of alkali-treated proteins and found 30-50% of ingested LAL was excreted in feces in bound form, probably representing unabsorbed material. Partial hydrolysis (to small peptides) of LAL containing proteins renders the LAL more available than if protein-bound, but less available than free LAL (DeGroot et al, 1976).

Not all "protein-bound" LAL is equivalent in availability or biological effect. Slump (1978) recently reported studies of enzymatic digestion of proteins containing various LAL levels. Enzymatic hydrolysis of an alkali-treated soy protein containing 13,000 ppm LAL resulted in the release of 3%, or 390 ppm LAL. Similar enzymatic hydrolysis of two samples of alkali-treated caseins which contained, respectively, 55,000 and 10,000 ppm LAL resulted in release of 0.2% and 0.5% (110 and 50 ppm) LAL as the free amino acid. When these three proteins were fed, nephrocytomegaly was not observed in rats fed a diet containing 10,000 ppm LAL bound in the alkali-treated casein which contained 55,000 ppm LAL, but was observed in rats fed both of the other proteins at much lower dietary LAL levels (25-2700 ppm) (Table 4). Finot (1977) has also reported a higher percentage of LAL released from proteins containing less LAL than from proteins containing very high amounts. In enzymatic hydrolysis studies of alkali-treated lactalbumins (16,000-18,000 ppm LAL) with pepsin-pancreatin combinations and continuous product removal via dialysis, he found that 27% of LAL was released from the protein.

These data indicate that the severity of alkaline treatment, type of protein and product concentration all contribute to the bioavailability of protein-bound LAL. The difference in LAL release in these enzymatic hydrolysis studies by Finot et al (1977) and Slump (1978) appears to be partly due to the

Table 4. Release of free LAL by enzymatic hydrolysis and occurrence of nephrocytomegaly in rats fed one sample of alkali-treated soybean protein and two samples of alkali-treated caseins for 4 weeks.[1]

Alkali-treated Protein	ppm LAL in Protein	ppm LAL Released by Enzymatic Hydrolysis	Protein-bound LAL in Rat Diet	Nephrocytomegaly[1]
Casein I	55,000	110	10,000	−
Casein II	10,000	50	2,700	+
Soy protein	13,000	390	2,500	+

[1]*Slump et al (1977).*
[2]−: *Absent;* +: *present.*

different hydrolysis times employed, suggesting that intestinal transit time may be a limiting factor in release of LAL. Steric hindrance of enzymic digestion by the LAL crosslink, due to, as suggested by Gould and MacGregor (1977) the degree and nature of the crosslink, may also be important in availability of LAL in various proteins.

Lysinoalanine has two asymmetric centers and, therefore, four possible stereoisomers. These have been recently synthesized and designated L-D, D-L, L-L, and D-D, the first part of the notation referring to the lysine, and the last, to the alanine moiety (Tas and Kleipool, 1976). Feeding studies have shown the L-D-LAL to be most toxic, causing some nephrocytomegaly at only 30 ppm. Next in potency is D-L (100 ppm), then L-L (300 ppm), and D-D is least potent (1,000 ppm) (Feron et al, 1977).

EXPERIMENTS WITH RADIOLABELED LYSINOALANINE

Metabolic studies with ^{14}C-LAL labeled in the lysine moiety have been conducted by Finot (1977, 1978), and recently by ourselves (Struthers et al, 1978c). Whole body autoradiography of rat, Swiss mouse, Syrian hamster, and Japanese quail, showed the highest ^{14}C concentration to be in kidneys in the rodents, but not in the quail. After eight hours, the rat kidney appeared uniformly labeled; by twenty-four hours, the label was concentrated in the outer stripe area in which neph-

rocytomegaly occurs (Finot et al, 1977). Administration of
^{14}C-LAL to rats either intraperitoneally or *per os* resulted in 10-25% of the total radioactivity being recovered as labeled CO_2 in the first twenty-four hours. After twenty-four
hours, approximately 3% of the total radioactivity was found
in the kidneys. Some of the ^{14}C label appeared to be bound
to proteins (Finot et al, 1977). Studies to determine 1) tissue distribution and rate of excretion of ^{14}C-LAL, and 2)
whether or not there are differences in rats pre-fed a control
diet or an LAL-containing diet for four weeks were recently
undertaken in our laboratory (Struthers et al, 1978c). All
rats were dosed by stomach tube with ^{14}C-LAL labeled in the
lysine moiety. Urine and feces were collected at 6, 12, 18,
24, 48, and 72 hours, and four rats from each group were sacrificed at each time period. Kidneys, liver, intestinal
tract, spleen, testes, and a blood sample were removed for
counting. (Preliminary studies with unlabeled LAL showed no
detectable LAL in other organs.) As expected, the main route
of excretion was in the urine. Figures 3 through 7 show excretion and tissue distribution of ^{14}C from ^{14}C-LAL. Only
a few significant differences were noted in the manner of handling LAL in the two groups of rats, but these are worth not-
ing. After six hours, the control rats had excreted 6%, and
the LAL pre-fed rats, 18% of the total dose (Figure 3). Figure 4 shows that the difference in early excretion rate was
due to urine excretion. Further, the kidneys of the controls
contained only 18% of the total dose after six hours, compared
to 30% of total dose in kidneys of the LAL-fed rats (Figure
5). By twelve hours, urinary and renal radioactivity was similar in the two groups of rats (Figures 4 and 5), but the intestinal tract of controls contained 19% of the total radioactivity, compared to 1% in the intestinal tract of the LAL-fed
group. By twenty-four hours, this also had equalized (Figure
6). Fecal excretion of radioactivity between twelve and eighteen hours was significantly higher in controls, and liver
^{14}C was signficantly higher at twenty-four hours in controls
than in the LAL-fed rats (Figure 7). These figures indicate
that there may be some metabolic adaptation to ingestion of
LAL.

The ^{14}C distribution was also examined in renal subcellular fractions separated by differential centrifugation at
twenty-four, forty-eight, and seventy-two hours after dosing
(Figure 8). There were some differences between control and
LAL-fed rats, particularly in the nuclei-cell debri fraction
at twenty-four and forty-eight hours. The highest level of activity in any fraction was found in the cytosol at twenty-four
hours in both groups. After this time, no subcellular fraction appeared to concentrate LAL to the exclusion of the other

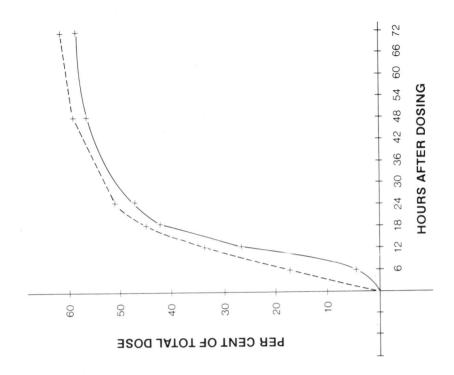

Figures 3 through 7: Excretion and tissue distribution of radioactivity from 14C-LAL (labeled in the lysine moiety) over a 72-hour period in control rats (——) or rats prefed an LAL-containing diet (----). Figure 3 - Accumulated total excretion in urine and feces. Figure 4 - Cumulative excretion (+) in urine, and amount excreted in urine at each sacrifice period (*). Figure 5 - Decrease of 14C in kidneys with time (controls, +———+; LAL-prefed, *----*). Figure 6 - 14C counts found in digestive tract (G), liver (L), and testes (T) with time. Figure 7 - 14C excreted in feces (cumulative, +; percent/time period, *).

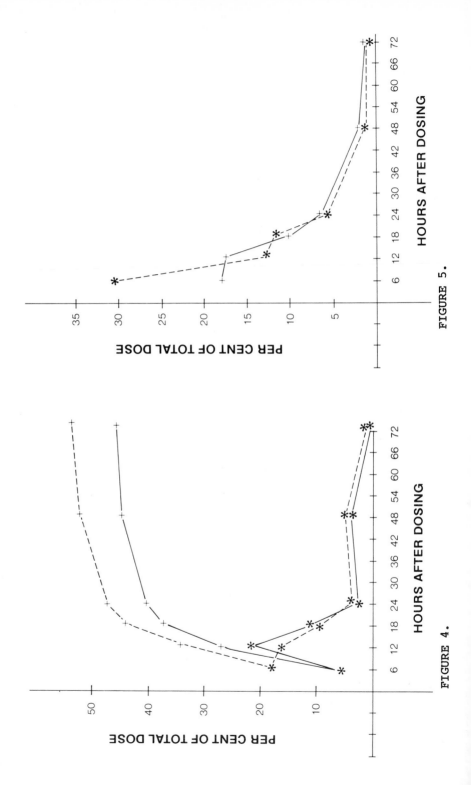

FIGURE 4.

FIGURE 5.

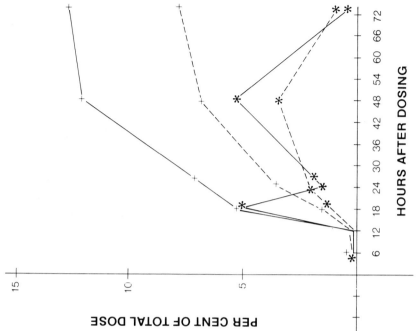

PER CENT OF TOTAL DOSE

HOURS AFTER DOSING

FIGURE 7.

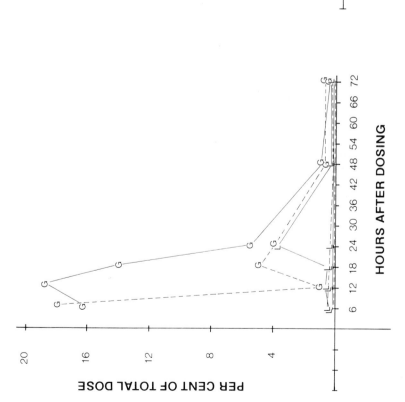

PER CENT OF TOTAL DOSE

HOURS AFTER DOSING

FIGURE 6.

fractions. Trichloroacetic acid precipitates of these fractions contained as much as 87% of the total activity in the
fraction (Figure 8). There was no precipitable radioactivity
in the soluble fraction after twenty-four hours. Hydrolysis
of the TCA precipitates followed by amino acid analysis
(Raymond & Viviano, 1977) and counting of the collected peaks
revealed that 30-80% of the label was in lysine and 0-10% in
LAL. This indicates that little, if any, LAL is covalently
bound in renal proteins (Struthers et al, 1978c).

When the pooled whole kidney homogenates from each sacrifice time were similarly analyzed, a major fraction of the ra-
diolabel was again found in lysine. Results are shown in Table 5. The ^{14}C-lysinoalanine used to dose the rats was 99.0%
LAL and 1.0% lysine, by both amino acid analysis and by determination of radioactivity in each peak. However, the lysine
in the kidneys accounted for up to 4.5-5% of the total ^{14}C
administered, i.e. there was more ^{14}C-lysine present in the
kidneys than there had been in the starting material. Apparently, the rat has some mechanism for converting a small fraction of ingested LAL to lysine, as well as to the several me-
tabolites found by Finot (1977). It seems reasonable to speculate that this is a renal enzyme, although there are numerous
other possibilities (intestinal bacteria, liver enzymes,
etc.). There is a further possibility, of course, and that is
that what we measured was an LAL metabolite which cochromato-
graphs with lysine in our system. A typical chromatograph is
shown in Figure 9.

Of considerable interest also was the fact that rats pre-
fed LAL excreted ^{14}C-LAL somewhat more rapidly than the control rats, despite the LAL-fed rats' continuing consumption of
a diet containing 3,000 ppm protein-bound LAL until they were
killed. In controls, renal radioactivity and renal LAL decreased throughout the experiment at parallel rates, whereas
in LAL-fed rats, renal radioactivity decreased in a manner
similar to controls, but the LAL level did not. Total renal
LAL in LAL-fed rats decreased between six and twelve hours to
a level of 250 ppm, at which point a plateau was reached, and
renal LAL remained at 150-250 ppm for the remainder of the experiment (Figure 10). The maintenance of a relatively constant renal LAL level and concomitant rapid excretion of the
^{14}C-LAL indicates a very rapid turnover rate for LAL in the
rat, and the presence of two separate LAL pools. (Struthers
et al, 1978c).

Total urinary and fecal excretion of LAL following a 260
mg dose was approximately 60% of dose in either control or
LAL-fed rats, seventy-two hours after administration
(Struthers et al, 1978c). Fifty percent was excreted in urine
and feces by twenty-four hours. It is known that 20-25% of

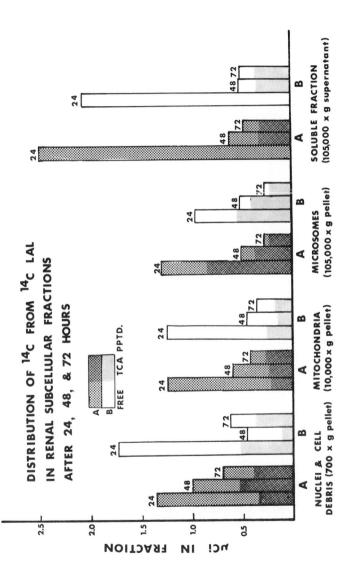

FIGURE 8. Distribution of label in renal subcellular fractions of rats prefed either control or LAL-containing diets sacrificed 24, 48, and 72 hours after a 260 mg oral dose of ^{14}C-LAL. The shaded part of each bar represents ^{14}C which is precipitable by trichloro-acetic acid (TCA) (Struthers et al, 1978c). A: Controls. B: Rats prefed LAL.

*Table 5.   Percent of total dose in LAL and Lysine in whole
        kidney homogenate following oral administration of
                    ^{14}C-LAL.[1]*

	Hrs. After Dosing	% of Dose Recovered As:	
		LAL	Lysine
Control (isolated soy protein)	6	10.57	5.05
	12	13.37	3.57
	18	6.07	2.04
	24	0.28	2.92
	48	0.29	0.86
	72	0.10	1.25
Pre-fed LAL (as alkali-treated soy protein)	6	19.07	4.55
	12	6.46	3.30
	18	6.24	3.51
	24	2.37	2.34
	48	0.10	0.98
	72	0.12	1.01

[1]*Struthers et al (1978c).*

the radioactivity from ^{14}C-LAL is excreted in expired air
within twenty-four hours (Finot, 1977). Therefore, the half-
life of ingested LAL must be less than twenty-four hours.

BIOLOGICAL SIGNIFICANCE OF LAL

Nephrocytomegalia precipitated by ingestion of LAL has for
some years been reported to occur only in the rat. One thou-

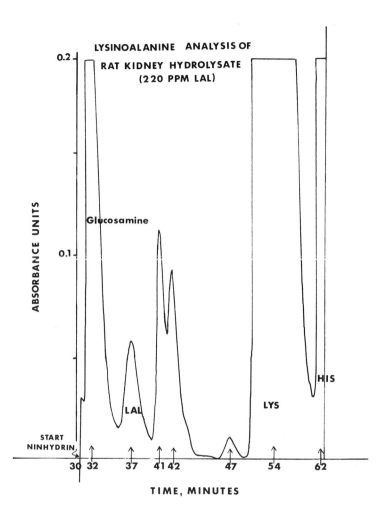

FIGURE 9. Chromatograph of kidney fraction. The kidney
homogenates were hydrolyzed in 6N HCl, and LAL analysis performed with Durrum D-500 amino acid analyzer (using DC4A sul-
fonated polystyrene resin in a 1.75 mm x 48 cm column, 300
psi). Amino acids were eluted with a single sodium citrate
buffer (.35N sodium, pH 5.30) at a column temperature of
65°C and flow rate of 7.9 ml/hr. A fraction collector connected to the photometer allowed peak collection directly into
scintillation vials. The neutral and acidic amino acids were
collected in one fraction, the basic amino acids separated as
shown in the above profile and counted as peaks (Struthers &
Raymond, 1978c).

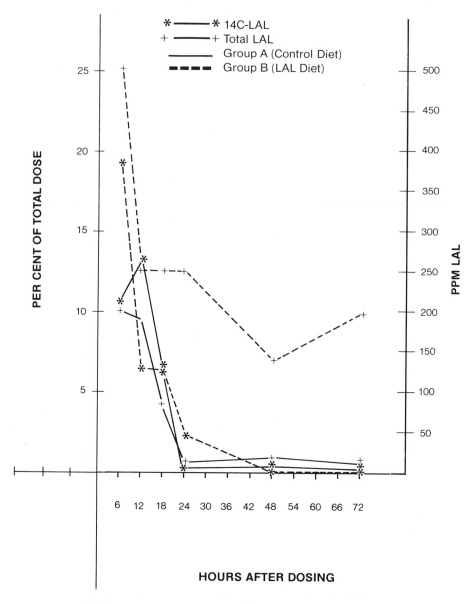

FIGURE 10. Concentration of LAL in rat kidney. Total LAL and ^{14}C-LAL decrease simultaneously in control rats; in LAL-fed rats, ^{14}C-LAL decreases throughout the experiment similar to controls, while total LAL decreases to around 200 ppm and plateaus. ^{14}C-LAL, **; total LAL, ++; Controls, _____; LAL-fed, ----.

sand ppm free LAL fed to Swiss mice, Syrian golden hamsters, dogs, Rhesus monkeys, rabbits, and Japanese quail for periods of four to nine weeks did not produce nephrocytomegaly (DeGroot et al, 1976). Recently, it has been reported that feeding 10,000 ppm free LAL to Swiss mice did induce a minimal degree of nephrocytomegaly; the same level of LAL fed to Syrian hamsters did not (Feron et al, 1977). Sternberg et al (1975) reported lysinoalanine content of a number of foods. We have also examined a number of foods, and results of this study are shown in Table 6. Nephrocytomegaly is not known to be a problem in humans despite the fact that some food materi- als which have been consumed for some years contain LAL. The nephrocytomegalic condition in rats appears to be largely re- versible if the LAL is removed from the diet, at least in short-term studies (Struthers et al, 1978a). In a sixteen- week study in which rats were fed a 3,000 ppm protein-bound LAL diet for eight weeks followed by a control diet for the subsequent eight weeks, the nephrocytomegaly score at sixteen weeks dropped to 30% of the score at eight weeks, and the in- cidence decreased from 100% to 38% of rats (Table 7). Feron (1977) has reported similar results.

Long-term effects of ingesting LAL or of nephrocytomegaly have been reported by only one group of investigators (Feron et al, 1978). A serial autopsy study was carried out on rats fed 0 to 200 ppm free synthetic LAL for one year. After four weeks, nephrocytomegaly was present: pale enlarged nuclei with a foamy chromatin pattern were seen, along with somewhat enlarged cells. During the course of the study, intranuclear inclusions and multiple nucleoli, such as have been reported when high doses of LAL are fed (Woodard, 1969, 1975), became more common, and exaggerated inequalities in cell and nuclear size became increasingly evident. Tubules in the affected intercortico-medullary area appeared somewhat disarranged. However, no signs of proliferative or neoplastic change in any of the kidneys was observed. Numerous carcinogenic agents which produce karyomegaly are known (reviewed by Gould and MacGregor, 1977). However, there is little similarity, except for the karyomegaly, between the LAL-induced lesion and vari- ous types of carcinogenic lesions (Feron, 1978).

Whether or not LAL had any teratogenic effects was of con- cern to us. We fed 0, 500, 1,000, 2,000, or 3,000 ppm protein-bound LAL to pregnant rats during gestation and lacta- tion (Struthers, et al, 1978b). No teratological effects were found, and no significant differences in conception rate, num- ber of young per litter, fetus weight, or mortality were ob- served (Table 8). In one of three experiments, pup birth weight was reduced significantly when the dams were fed a diet containing 3,000 ppm LAL (in which the alkali-treated soy pro-

Table 6.   LAL in some commerical products.[1,2]

Product	LAL in Sample, ppm	LAL in Protein, ppm
Casein		
from non-pasteurized		
whole milk	0	0
Sodium caseinate	270	310
Processed Cereals		
5 samples	0	0
Oats	25	160
Rice	70	1,000
Pasta and Bread		
4 samples	0	0
Cheese		
4 samples	0	0
Canned Fruits and Vegetables		
5 samples	0	0
Snack Foods		
7 samples	0	0
Pretzels	20	220
Egg		
Spray dried whole	0	0
Boiled 30 minutes		
white	0	0
yolk	0	0
Milk		
Evaporated, sample 1	50	200
sample 2	140	550
Dry Powder	0	0
Luncheon meat		
5 samples	0	0
Canned Fish		
5 samples	0	0
Sardines	120	270
Strained Beef for Infants	120	160

[1]Method of Raymond and Viviano, 1977.

[2]Values reported represent samples of several brands of commercial products analyzed between 12/75 and 10/76. Since only one sample of each product was analyzed, values may not be typical.

Table 7. *Reversibility of LAL-caused nephrocytomegaly upon removal of LAL from the diet.*[1]

| | Cytomegaly[3] | | | |
| | 8 Weeks | | 16 Weeks[4] | |
Diet	Score[3,6]	Incidence	Score[3,6]	Incidence
MALES				
A - Ad Lib Control	-	-	0.0a	0/10
B - Pair Fed Control	0.0a	0/2	0.0a	0/8
C - ATSP-Fed[2]	2.0b	2/2	0.78b	4/8
FEMALES				
A - Ad Lib Control	-	-	0.00	0/10
B - Pair Fed Control	0.0a	0/2	0.0a	0/8
C - ATSP-Fed[2]	1.5	2/2	0.33b	2/8
SEM[5]	0.33	-	0.06	-

[1]Data from Struthers et al (1978).
[2]ATSP diet: 30% of an alkali-treated soy protein providing 3,000 ppm LAL.
[3]Score: cytomegaly was scored on a 0-3 (absent-severe) scale.
[4]Experimental rats were fed the ATSP diet for the first 8 weeks, and switched to the control diet for the second 8 weeks.
[5]SEM: standard error of mean.
[6]Means not followed by the same letter are significantly different, P ≤.05 (Duncan's Multiple Range Test, Steele and Torrie, 1960).

tein was the sole protein source) both prior to conception and during gestation. In the other two experiments, timed pregnant females were purchased, and no significant difference in birth weight was seen. However, in all the experiments, pup weight gain was significantly reduced when dams were fed diets containing 20% or 30% alkali-treated soy protein (supplying 2,000 or 3,000 ppm LAL) (Table 9). Pups from dams fed 3,000

Table 8. Summary of data from a teratalogical study with rats[1],[2] fed a diet containing 30% of either a soybean control protein or an alkali-treated soybean protein (ATSP)

	Control	ATSP
Corpora Lutea	9.2 ± 0.84	9.7 ± 2.31
Implants	7.8 ± 1.1	8.3 ± 5.51
Resorption Sites	1.0 ± 1.0	1.3 ± 1.53
No. Fetuses/Litter	7.8 ± 1.1	8.3 ± 5.51
Fetus Wt., g	3.15 ± 0.62	2.48 ± 0.23
Crown-rump Length, cm	3.7 ± 0.19	3.4 ± 0.85

[1]*Struthers et al (1978b).*
[2]*Diets contained 30% control or alkali-treated soy protein, the latter diet supplying 3,000 ppm LAL.*

ppm LAL also developed nephrocytomegaly by thirty days of age. It was not known whether the nephrocytomegaly in the pups was due to their consumption of the dams' diet, or to LAL in the milk. Analysis of the milk proved negative for LAL, and there were no other apparent compositional changes. It was concluded that the reduced weight gains were the result of decreased milk production caused by the dams' consumption of an inferior protein, rather than to LAL per se. In all of our experiments, we have added amino acids to meet NRC requirements (Nutrient..., 1972) and to match the amino acid composition of the control soy protein. Despite supplementation, we find the PER for the supplemented 3,000 ppm LAL diet (30% alkali-treated soy protein) to be only 1.8, compared to 2.8 for the 30% soybean protein control diet (values adjusted to 2.5 for casein). It appears that proteins crosslinked by LAL, similar to such naturally crosslinked proteins as keratin and collagen, are poorly digestible. Amounts of dietary protein-bound LAL at levels lower than 2,000 ppm were not found to have any effect in rats. Data showing similar nutritional quality reduction has been previously reported (Van Beek et al, 1974).

Dietary LAL in products currently on the market, particularly in view of the varied diet consumed by humans, seems to

Table 9. *Birth and sacrifice weights of rat pups from dams*
fed 0-3,000 ppm protein-bound LAL.[2]

Dietary LAL, Ppm	0	500	1,000	2,000	3,000	SEM[3]
Pup birth wt., g						
Expt. #1	6.60a[1]	–	–	–	6.10b	0.14
2	6.46a	6.25a	6.38a	6.00a	–	0.20
3	6.80a	–	6.24a	6.03a	5.82a	0.47
Pup sacrifice wt., g						
14 days	24.72a	–	16.82a	15.63a	18.55a	3.00
17 days	40.00a	32.00a	34.50a	27.00b	–	5.01
29 days	88.00a	91.00a	88.00a	72.67b	–	11.27
30 days	99.80a	–	–	–	79.30b	9.78

[1]*Means not followed by same letter are significantly*
different (P $\leq$.05). Duncan's Multiple Range Test (Steele and
Torrie (1960).
[2]*Struthers et al (1978).*
[3]*SEM: standard error of mean.*

represent no health hazard. Perhaps the greatest problem
which could be associated with LAL would be the reduced nutritional value of protein products which contained relatively
high levels of LAL, and these would be significant only if
consumed as a major portion of dietary protein.

ACKNOWLEDGEMENTS

The authors wish to thank Dr. Paul Finot, the Nestle' Company,
for prepublication copies and English translations of work in
his laboratory, and Dr. John Finley, USDA Western Regional
Labs, for supplying us with a prepublication copy of his method for LAL synthesis.

REFERENCES

DeGroot, A. P. and Slump, P. (1969). J. Nutr. 98: 45-56.
DeGroot, A. P.; Slump, P.; Feron, V. J.; and Van Beek, L.
(1976). J. Nutr. 106: 1527-1538.

DeGroot, A. P.; Slump, P.; Van Beek L.; and Feron, V. J. (1977). *Proteins for Humans: Evaluation and factors affecting nutritional value* (Bodwell, C. E., ed.). AVI, Westport, CT, pp. 270-283.

Feeney, R. E. (1977). *Proteins for Humans: Evaluation and Factors affecting nutritional value* (Bodwell, C. E., ed.). AVI, Westport, CT, Ch. 11, 233-254.

Feron, V. J.; Van Beek, L.; Slump, P.; and Beems, R. B. (1978). *Biochemical aspects of New Protein Food* (Alder-Nisen, J. ed.), Vol. 44, Proceedings of the 11th FEBS, Copenhagen. Pergamon Press, pp. 139-147.

Finot, Paul-Andre; Bujard, Elaine; and Arnaud, Maurice (1977). Adv. Exp. Biol Med. 86B: 51-71.

Finot, P.A.; Magnenat, Edith; Mottu, Francoise, and Bujard, Elaine; with the technical assistance of Deutsch, R.; Dormond, C.; Isely, A.; Pignat, L.; and Madelaine, R. (1978). Annales de la Nutrition et de l'Alimentation. In press.

Gould, D. H. and MacGregor, J. T. (1977). Adv. Exp. Med. Biol. 86B: 29-48.

Gross, E.; Chen, H. C.; and Brown, J. H. (1973). Peptides with Lanthionines and x, β-unsaturated amino acids: cinnamycin and duramycin. Abstracts of 166th meeting of the American Chemical Society Biol. 80.

Hodgins, D. S. (1971). J. Biol. Chem. 246: 2977-85.

Horne, Millard J.; Jones, D. Breese; Ringel, S. J. (1941). J. Biol. Chem. 138: 141-149.

Kaunitz, H. and Johnson, R. E. (1976). Metab. Clin. Exp. 25: 69-77.

Newberne, P. M. and Young, V. R. (1966). J. Nutr. 89: 69-79.

Nutrient Requirements of the Laboratory Rat. (1972) *Nutrient Requirements of Laboratory Animals,* second revised edition. National Academy of Sciences, publication No. ISBN 0-309-0208-X. Washington, DC.

Patchornik, Abraham and Sokolovsky, Mordechai (1964). J. Am. Chem. Soc. 85: 1860-61.

Prescher, E. E. and Struthers, B. J., Unpublished Observations. (1975-78).

Raymond, M. L. and Viviano, D. A., A rapid and sensitive analysis for lysinoalanine in proteins and its application to food products. Presented at the 174th ACS meeting. Ag. Fd. Abs. 73, Chicago (1977).

Reyniers, Jon P.; Woodard, James C.; and Alvarez, Marvin R. (1974). Lab. Invest. 30: 582-588.

Slump, P., with the technical assistance of Jongerius, G. C.; Kraaikamp, W. C.; and Schrueder, H. A. W. (1978). Lysino-alanine in alkali-treated proteins and factors influencing its biological activity. Annales de la Nutrition et de l'Alimentation. In Press.

Steele, R. G. D. and Torrie, J. H. (1960). *Principles and Procedures of Statistics.* McGraw-Hill, New York, pp. 16-18, 106-109.

Sternberg, M.; Kim, C. Y.; and Schwende, F. J. (1975). Science 190: 992-994.

Struthers, Barbara J.; Dahlgren, Robert R.; and Hopkins, Daniel T. (1977). J. Nutr. 107: 1190-1198.

Struthers, Barbara J.; Hopkins, Daniel T.; and Dahlgren, Robert R. (1978a). J. Fd. Sci. 43: 616-618.

Struthers, Barbara J.; Hopkins, Daniel T.; Prescher, Elmer E.; and Dahlgren, Robert R. (1978b). J. Nutr. 108: 954-958.

Struthers, B. J.; Brielmaier; J. R.; Raymond, M. L.; and Hopkins, D. T. (1978c). Fed. Amer. Soc. Exp. Biol., 62nd ann. meeting. Abs. 1540.

Tas, A. C. and Kleipool, R. J. C. (1976). Lebensm. Wiss. U. Technol. 9: 360-362.

Van Beek, L.; Feron, V. J.; and deGroot, A. P. (1974). J. Nutr. 104: 1630-1636.

Woodard, James C. and Alvarez, Marvin R. (1967). Arch. Path. 84: 153-162.

Woodard, James C. (1969). Laboratory Invest. 20: 9-16.

Woodard, James C. (1971). Amer. J. Path. 65: 269-278.

Woodard, J. C. (1975). Vet. Pathol. 12, 65-66.

Woodard, James C. and Short, Dennis D. (1973). J. Nutr. 103: 569-574.

Woodard, J. C.; Short, D. D.; Strattan, C. E.; and Duncan, J. H. (1977). Fd. Cosmet. Toxicol. 15: 109-115.

Woodard, J. C. and Short, D. D. (1977). Fd. Cosmet. Toxicol. 15: 117-119.

Woodard, J. C.; Short, D. D.; Alvarez, M. R.; and Reyniers, J. (1975). *Protein Nutritional Quality of Foods and Feeds II:* (Friedmann M., ed.) 595-618, Marcel Dekker, NY.

Ziegler, K.; Melchert, J.; and Lurken, C. (1967). Nature 214: 404-405.

DISCUSSION

Erdman: All of the toxicological studies reported in your paper were performed with intact, healthy animals. I am curious whether anyone has investigated the effects of feeding Lysinoalanine to animals with some sort of kidney disorder, such as necrosis. The stress of kidney malfunction could have a synergistic effect upon the toxicity of LAL and could have implications for persons with reduced kidney function.

Struthers: I don't think anyone is going to be able to
answer that particular question until we find the biochemical
lesion which causes the observed nephrocytomegalic effects,
and we are still looking for this. You're right, there might
be some problems; I don't know. It appears that LAL is largely handled by the kidneys as a foreign compound, and anytime
that the kidneys are damaged or the kidney function is altered, altered excretion of other kinds of foreign compounds
also occurs.

VEGETABLE PROTEIN AND LIPID METABOLISM

K. K. Carroll, M. W. Huff and D. C. K. Roberts

The common method of producing atherosclerotic lesions in
experimental animals is to feed cholesterol, but it has been
known for some time that hypercholesterolemia and atherosclerosis can be produced in rabbits by feeding semipurified diets
containing little or no cholesterol. Because these effects
could be largely prevented by including substantial amounts of
polyunsaturated fat in the semipurified diets, it was suggested that they were due to a relative deficiency of essential
fatty acids. However, rabbits fed commercial diets do not
normally develop hypercholesterolemia and atherosclerosis, and
it has been shown that this difference between commercial and
semipurified diets is not due to the small amount of polyunsaturated fat in commercial diets (Carroll and Hamilton, 1975;
Carroll, 1978).

The differing effects of commercial and semipurified diets
on the level of plasma cholesterol in rabbits were confirmed
in our laboratory (Carroll, 1971), and further experiments
were initiated to determine the reasons for the difference.
In these experiments, systematic variation of the composition
of a low fat, cholesterol-free, semipurified diet showed that
the hypercholesterolemic response was due to the casein used
as dietary protein (Hamilton and Carroll, 1976). When the casein in the semipurified diet was replaced by isolated soy
protein, the level of plasma cholesterol remained low, as in
rabbits on commercial diet (Fig. 1). Feeding trials with
semipurified diets containing proteins from a variety of
sources showed that animal proteins generally gave a hypercholesterolemic response, whereas plant proteins consistently
gave low levels of plasma cholesterol. These results are summarized in Table 1. It can be seen that some diets gave better growth than others, but there seemed to be no correlation
between growth rate and level of plasma cholesterol.

Feeding trials with mixtures of casein and soy protein
isolate (Huff, et al, 1977a) showed that a relatively high

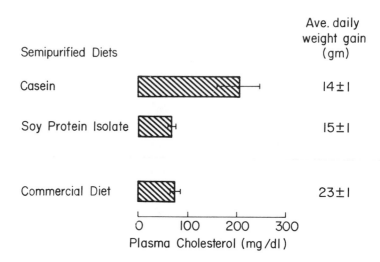

FIGURE 1. Plasma cholesterol levels and weight gains in rabbits fed low fat, cholesterol-free, semipurified diets or a commercial diet (Huff, et al, 1977a). Results are given as mean±SEM for groups of 6 rabbits fed the diets for 28 days.

proportion of casein was required to produce a hypercholester-olemic response (Fig. 2). Thus, a 1:1 mixture gave a low lev-el of plasma cholesterol comparable to that obtained when soy protein isolate was used alone. A 3:1 mixture produced some elevation of plasma cholesterol, but not as much as casein by itself. The protein mixtures gave better weight gains than either protein alone (Fig. 2).

It is of interest to know whether the differing effects on plasma cholesterol are due to differences in amino acid compo-sition of the proteins or to other components of the protein preparations. It seems unlikely that the results shown in Table 1 were influenced by dietary fat, since most of the fat was removed from the protein preparations used for these ex-periments, and the basal diet contained only enough fat to provide a supply of essential fatty acids. However, many of the protein preparations, particularly those from plant sources, were concentrates which contained 30-40% of non-protein material. Although the diets were made isonitrogenous

Table 1. Effects on Growth Rate and Plasma Cholesterol Level of Rabbits Fed Various Protein
          Preparations in Low Fat, Cholesterol-free Diets[a]

Animal Proteins				Plant Proteins			
Protein	Number of animals	Weight gain (g/day)	Plasma cholesterol (mg/dl)	Protein	Number of animals	Weight gain (g/day)	Plasma cholesterol (mg/dl)
Egg yolk	5	11±2	268±35	Detoxified rapeseed flour	6	6±2	99±11
Skim milk	6	18±3	230±40	Wheat gluten	6	3±1	80±21
Turkey	6	13±3	219±30	Peanut	6	15±2	80±10
Lactalbumen	5	9±2	215±69	Oat	6	10±2	77±14
Casein	6	14±1	204±44	Cottonseed	6	17±4	76±14
Whole egg	6	2±3	183±26	Sesame seed	6	18±2	70±5
Fish	6	14±3	166±32	Alfalfa	4	14±6	69±13
Beef	5	20±2	160±60	Soy protein isolate	6	14±2	67±9
Chicken	5	15±3	149±18	Sunflower seed	6	15±2	53±12
Pork	6	25±1	110±17	Pea	6	7±2	41±11
Egg white	6	9±2	105±28	Faba bean	6	8±2	30±4

[a]Most of these results have been reported previously (Hamilton and Carroll, 1976; Huff, et al, 1977a). The experiments with egg yolk, whole egg, turkey and chicken were carried out by Anne Weisz. These were delipided as described by Hamilton and Carroll (1976). Senad Kosaric carried out the experiment with alfalfa, which was delipided by extraction with chloroform-methanol (2:1) and then with methanol. The whole egg powder was kindly donated by W. E. Melby, Viobin Corporation, Monticello, Illinois and the alfalfa (whole leaf protein-Pro-Xan I) by G. O. Kohler, Western Regional Research Laboratory, U.S.D.A., Berkeley, California.

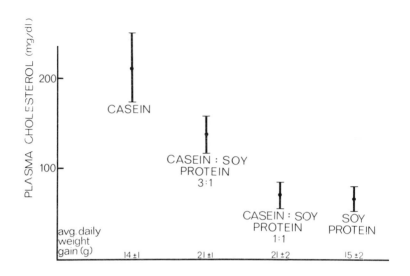

FIGURE 2. Plasma cholesterol levels and weight gains in rabbits fed low fat, cholesterol-free, semipurified diets containing mixtures of casein and soy protein isolate. Results are given as mean+SEM for groups of 5 or 6 rabbits fed the diets for 28 days.

in each case by reducing the level of dextrose to allow for non-nitrogenous components of the protein preparations, these components may still have affected the results obtained.

Experiments with diets containing protein hydrolysates or mixtures of L-amino acids have provided evidence that at least part of the difference in the effects of dietary proteins on plasma cholesterol is due to differences in their amino acid composition (Huff, et al, 1977a). An enzymatic digest of casein, or a mixture of L-amino acids equivalent to casein gave hypercholesterolemic responses similar to those obtained with casein itself. An enzymatic digest of soy protein gave a low level of plasma cholesterol comparable to that obtained with soy protein isolate, but a mixture of L-amino acids equivalent to soy protein gave a somewhat higher level of plasma cholesterol (Fig. 3). This may mean that the hypocholesterolemic properties of soy protein isolate are partly due to factors other than the protein itself. However, feeding a mixture of amino acids is not equivalent to feeding a protein containing amino acids in the same proportions, since the protein must be digested before its amino acids are available for absorption.

Further attempts to investigate this problem have not so far yielded clear-cut results. Supplementing the soy protein diet with methionine (Hamilton and Carroll, 1976), or increasing the proportion of essential amino acids relative to nonessential amino acids (Huff, et al, 1977) did not seem to have

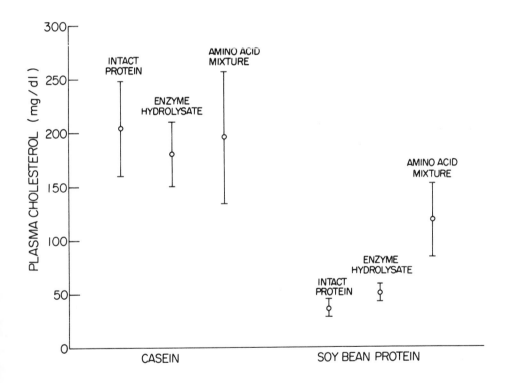

FIGURE 3. Plasma cholesterol levels of rabbits fed diets containing protein hydrolysates or mixtures of amino acids equivalent to casein or soy protein isolate. Results are given as mean+SEM for groups of 6 rabbits fed the diets for 28 days. (Reproduced from Carroll, et al, 1977, by permission of the publisher.)

much effect on the level of plasma cholesterol. An attempt
was also made to reverse the effects of casein and soy protein
isolate by adding amino acids to casein (approximately 1:1) to
give a mixture with the overall amino acid composition of soy
protein, and by adding amino acids to soy protein isolate to
give a mixture corresponding to casein. However, the added
amino acids appeared to have little influence on the results
obtained (Huff, et al, 1977b).

Long-term feeding of a semipurified diet containing casein
has shown that the hypercholesterolemia persists for periods
of as long as 10 months, whereas the plasma cholesterol remains low in rabbits fed a corresponding diet containing soy
protein isolate (Carroll, et al, 1977) (Fig. 4). Growth rates

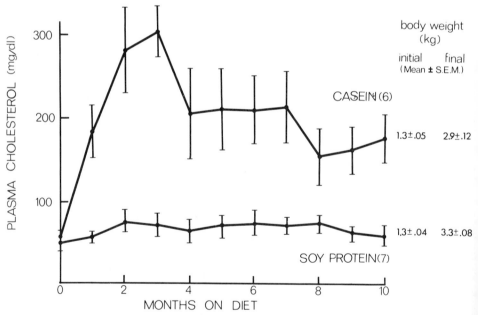

FIGURE 4. Plasma cholesterol levels of rabbits fed low
fat, cholesterol-free diets for 10 months. The numbers of
rabbits fed diets containing casein or soy protein isolate are
shown in parentheses. (Reproduced from Carroll, et al, 1977,
by permission of the publisher.)

were about the same on both diets. Rabbits autopsied after 10 months on the casein diet showed varying degrees of aortic atherosclerosis, whereas few if any lesions were seen in rabbits fed the soy protein diet for the same length of time (Fig. 5).

Analysis of the plasma lipoproteins after the animals had been on these diets for about 5 months showed that the excess plasma cholesterol in rabbits on the casein diet was mainly in the intermediate density lipoproteins (IDL-d 1.006-1.019) although there was some increase in the very low density (VLDL-d < 1.006) and low density (LDL-d 1.019-1.063) lipoproteins (Fig. 6). The phospholipid and protein components of these fractions were also increased relative to those in lipoproteins from rabbits fed the soy protein diet, but no changes were observed in the triglyceride component. Plasma triglycerides were low in both dietary groups, perhaps because of low levels of dietary fat.

Turnover rates of cholesterol and the size of body pools of cholesterol were determined in rabbits on these two diets by injecting [26-^{14}C] cholesterol and analyzing the die-away curves of plasma cholesterol radioactivity, as described by Goodman and Noble (1968) and Nestel, et al (1969). The results (Huff and Carroll, 1977) showed that turnover was faster and the body pools were smaller in animals on the soy protein diet (Fig. 7). Analysis of fecal neutral steroids and bile acids also showed that both were excreted in larger amounts by rabbits on this diet (Huff and Carroll, unpublished experiments). Fumagalli, et al, (1978) have recently reported that replacement of casein by soybean meal in a semipurified diet fed to rabbits caused an increase in fecal excretion of neutral steroids, but not of bile acids. It seems possible that the lower levels of plasma cholesterol in rabbits on soy protein diets may be due to a higher rate of removal of cholesterol from the body pools, compared to rabbits on casein diets.

Experiments have also been carried out to see whether the level of plasma cholesterol in human subjects is influenced by replacing animal protein in the diet by soybean protein (Carroll, et al, 1978). These showed a small, but statistically significant, difference in the plasma cholesterol of young female students consuming either a diet in which 70% of the protein was derived from meat and milk, or a corresponding diet in which the meat and milk were replaced by protein meat analogues and soy milk (Fig. 8). Both diets were similar in fat and sterol content.

Much larger differences have been reported in similar experiments on Type II hypercholesterolemic patients by Sirtori, et al (1977). In their experiments, the observed hypocholesterolemic effect of replacing animal protein in the diet by

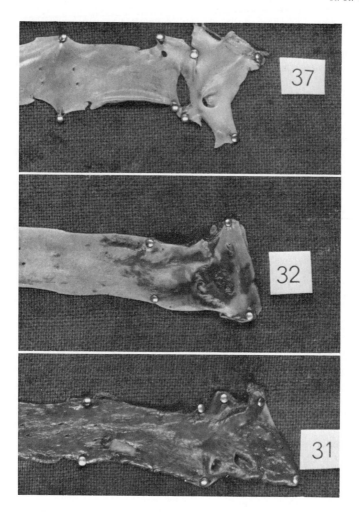

FIGURE 5. Aortas from rabbits fed low fat, cholesterol-
free diets containing either casein (31 and 32) or soy protein
isolate (37) for a period of 10 months. The aortas were
stained with Sudan III. Those from the casein group were chosen to show the extent of variation of lesions in animals of
this group. The aortas from all animals of the group fed soy
protein isolate were similar in appearance to the example
shown.

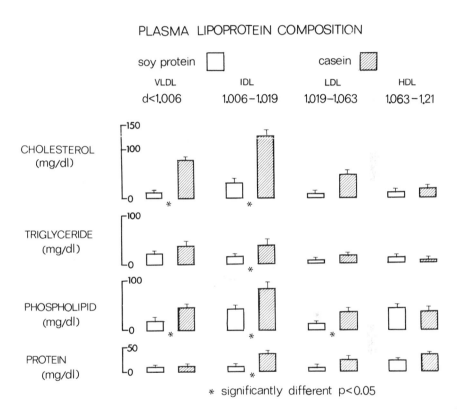

FIGURE 6. Analysis of plasma lipoproteins of rabbits after 5 months on low fat, cholesterol-free diets containing either soy protein isolate or casein. Results are expressed as
mean±SEM for groups of 4 rabbits. The lipoproteins were separated ultracentrifugally by the method of Redgrave, et al
(1975).

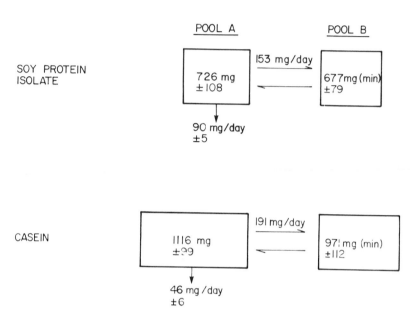

FIGURE 7. Two-pool analysis of cholesterol metabolism in rabbits fed low fat, cholesterol-free diets containing either soy protein isolate or casein. Results are expressed as mean+SEM for groups of 6 rabbits.

soybean protein was in addition to that obtained by feeding a low fat, polyunsaturated diet and did not appear to be altered by the addition of 500 mg/day of cholesterol to this diet. Patients used in their experiments had average cholesterol levels of greater than 300 mg/dl, and this may have been why the hypocholesterolemic effects were much greater than those observed in our studies on normocholesterolemic subjects. More experiments are needed to confirm and extend these findings.

In the above experiments with human subjects, the lowering of plasma cholesterol was achieved by replacing essentially all of the animal protein in the diet by soybean protein. However, in our studies with rabbits, it was found that the plasma cholesterol remained low in animals fed a diet containing a 1:1 mixture of animal and plant protein (Fig. 2). If this result can be extrapolated to humans, it suggests that a significant lowering of plasma cholesterol might be achieved

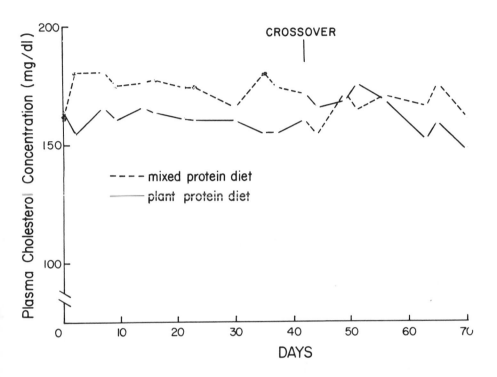

FIGURE 8. Plasma cholesterol levels in young female subjects consuming diets containing either a mixture of animal and plant proteins, or plant proteins only. (Reproduced from Carroll, et al, 1978, by permission of the publisher.)

without eliminating animal protein completely from the diet. Some adjustment in the relative proportions of animal and plant protein may be sufficient.

According to Kritchevsky (1977), the nutrient availability of animal protein relative to vegetable protein has more than doubled in the United States during this century from a ratio of 1.06 in 1909, to 2.37 in 1972. It seems possible that this may have played a significant role in the apparent increase in cardiovascular disease in the American population during this period. It is noteworthy that animal protein shows a higher positive correlation with mortality from coronary heart dis-

ease in different parts of the world than any other dietary
variable (Connor and Connor, 1972).

Diet has long been considered to play an important role in
hypercholesterolemia and atherosclerosis, but most of the emphasis has been on dietary fat and cholesterol. Evidence presented in this brief review suggests that more consideration
should be given to dietary protein.

ACKNOWLEDGEMENT

Support of this work by the Ontario Heart Foundation and
the Medical Research Council of Canada is greatfully
acknowledged. Dr. Carroll is an Associate of the Medical
Research Council of Canada, Mr. Huff is the recipient of
Ontario Graduate Fellowship and Dr. Roberts is the recipient
of Fellowships from the Medical Research Council and the
Ontario Heart Foundation.

REFERENCES

Carroll, K. K. (1971). *Atherosclerosis. 13*, 67-76.

Carroll, K. K. (1978). *Nutr. Rev. 36*, 1-5.

Carroll, K. K.; Giovannetti, P. M.; Huff, M. W.;
Moase, O.; Roberts, D. C. K.; and Wolfe, B. M. (1978). *Amer.
J. Clin. Nutr.* (in press).

Carroll, K. K., and Hamilton, R. M. G. (1975). *J. Food
Sci. 40*, 18-23.

Carroll, K. K.; Huff, M. W.; and Roberts, D. C. K.
(1977). In "Atherosclerosis IV. Proceedings of the Fourth
International Symposium." (G. Schettler, Y. Goto, Y. Hata and
G. Klose, eds.). pp. 445-448, Springer-Verlag, Berlin.

Connor, W. E., and Connor, S. L. (1972). *Prev. Med. 1*,
49-83.

Fumagalli, R.; Paoletti, R.; and Howard, A. N. (1978).
Life Sci. 22, 947-952.

Goodman, D. S., and Noble, R. P. (1968). *J. Clin.
Invest. 47*, 231-241.

Hamilton, R. M. G., and Carroll, K. K. (1976).
Atherosclerosis. 24, 47-62.

Huff, M. W., and Carroll, K. K. (1977). *Fed. Proc. 36*,
1104.

Huff, M. W.; Hamilton, R. M. G.; and Carroll, K. K.
(1977a). *Atherosclerosis. 28*, 187-195.

Huff, M. W.; Hamilton, R. M. G.; and Carroll, K. K.
(1977b). Adv. Exp. Biol. Med., Vol. 82, "Atherosclerosis:

Metabolic, Morphologic and Clinical Aspects." (G. W. Manning and M. D. Haust, eds.). pp. 275-277, Plenum Press, New York.

Kritchevsky, D. (1977). *Lipids. 12*, 49-52.

Nestel, P. J.; Whyte, H. M.; and Goodman, D. S. (1969). *J. Clin. Invest. 48*, 982-991.

Redgrave, T. G.; Roberts, D. C. K.; and West, C. E. (1975). *Anal. Biochem. 65*, 42-49.

Sirtori, C. R.; Agradi, E.; Conti, F.; Mantero, O.; and Gatti, E. (1977). *Lancet. 1*, 275-277.

DISCUSSION

Rosenfield: Dr. Carroll, with reference to your slide which showed cholesterol levels of rabbits on casein diets going up and then down to lower levels found at the initiation of the experiment, we have found similar results in experiments underway in our laboratory with Rhesus monkeys. After four months of feeding a semi-purified casein diet, we observed a rise in total cholesterol from 206 mg/dl to 271; now after eight months of feeding, it's back down to around 200. Interestingly, the ration of α-cholesterol (HDL) to β-cholesterol (LDL) has shown a decrease which remains significant after eight months, though not as low as in prior months.

Carroll: That's very interesting. I might just comment that we have done experiments in which we fed rabbits first on a casein diet, and then on a soy protein diet, or vice versa. The plasma cholesterol levels increased on the casein diet, and then came back down again when the casein was replaced by soy protein. When we started with the soy protein diet and then changed to the casein diet, the plasma cholesterol levels rose after the changeover.

Bodwell: I gather that all of these studies were done with fairly low lipid diets?

Carroll: Yes, except that I showed one slide in which the diets contained 15% butter or 15% corn oil. We have just recently done some further experiments in which we reduced the protein content to 20% and added 5% corn oil instead of 1%. I think the hypercholesterolemia was a bit less than we got with the 1% corn oil diet, but not appreciably so.

Harper: You could account for a high proportion of the total protein of the products from the amino acid analyses. What were the major differences in amino acid content between casein and the soy protein isolate?

Carroll: I have a couple of slides of results taken from
the literature showing comparisions between the essential and
non-essential acids of casein and soy protein isolate (see also Table 1 in Huff et al, 1977a). The major differences in
essential amino acids are in arginine and methionine, otherwise the amounts are much the same. Among the non-essential
amino acids, one thinks of glutamic acid as predominating in
plant proteins more than in animal proteins; but casein, in
fact, contains almost as much as soy protein. However, there
are differences in aspartic acid and proline, as well as alanine, cystine and glycine. We have certainly tried to look at
the problem from this point of view. In fact, Dr. David
Roberts prepared a computer program from the analytical data
for all the proteins that we had fed, and tried to come up
with a formula that might give a correlation between amino acid composition and plasma cholesterol levels. He did find a
combination which showed a correlation of about 0.8 for the
range of proteins I showed on an earlier slide, but the correlation was not applicable to mixtures of amino acids that we
fed. However, I don't think that you can look at the amino
acid composition of intact proteins and feel that they will be
absorbed from the gut in the same proportions during digestion. Although we have done a number of experiments with L-
amino acids in different combinations, I am never quite sure
whether this is a realistic approach to the problem; but I do
feel we have shown that we can alter plasma cholesterol levels
by feeding different combinations of amino acids.

Schonfeld: This is a fascinating series of studies, and
for those of us who are interested in lipoproteins it's a
whole new world. While there are several questions that could
be raised, I would like to make one comment and then ask a
couple of questions. The comment is about the relative ath-
erogenicity of the animal protein diet. It's hard to know exactly why this diet is more atherogenic than the soy protein
diet, but I suspect that any substance or group of substances
or any other manipulation which will raise the lipoprotein
levels in plasma will result in atherogenesis at the tissue
level, because the residence time of the lipoprotein in plasma
will be increased and because the gradient of lipoproteins between plasma and tissues will also be increased. So I'm not
sure that there's any specific atherogenic effect attributable
to this particular diet. In other words, from the point of
view of the susceptible tissues, the diet may be just another
of several ways of raising plasma lipoprotein levels, and this
is followed by atherogenesis.

    I didn't clearly understand your proposed mechanism for
the lipid effects of the diets. Did you say that neutral ste-

roid and bile excretion in the feces was increased? By which
diet?

   Carroll: By the soy protein diet.
   We've analyzed these by GLC and by isotopic dilutions and
they, if my memory is correct both the bile acids and the neutral steroids, are increased on the soy diet in the order of
being doubled. So that fits with the analysis of the curves
where it shows more sterol coming out of pool A on the soy
protein diet relative to the casein diet.

   Schonfeld: So one of the possible mechanisms would then
be that bile acid production and cholesterol excretion through
the liver are increased by the soy protein diet?

   Carroll: Yes, I think there is a real possibility that
cholesterol is being taken out of the system and not being replaced at quite the same rate, giving rise to a new steady
state level of plasma cholesterol.

   Schonfeld: Do you have any data on the synthesis of hepatic lipoproteins?

   Carroll: We haven't studied synthesis. We are trying to
get some studies going with lipoproteins to see where the rate
of turnover is altered and whether there is a blockage at some
point. We have studied acetate incorporation into cholesterol
in the liver (Carroll, 1971). I guess that if the turnover of
the whole system is slowed down, you would see a decreased
production of lipoproteins by the liver, but this may not be
the primary effect. The various steps in lipoprotein metabolism are operating as a sequence, and it is difficult to determine exactly where this may be influenced by the dietary
protein.

   Schonfeld: One final question. Do you have any data on
the effects of these various proteins on the secretion of the
gut hormones and of insulin.

   Carroll: No, we haven't; and I think that would be a
worthwhile thing to explore. One wonders where trypsin inhibitors, for example, might be having some influence; but I
don't really know whether this could explain the hypocholesterolemic effects of all the different plant proteins tested.
I don't have enough information about the existence of trypsin
inhibitors in different plant proteins.

   Schonfeld: Well, it's difficult to understand how the
lipoproteins are accumulating in plasma unless you increased

their production; and one way that I can envision that happening is by some hormonal mechanism perhaps due to different stimulatory effects of the amino acid mixtures in the two diets on insulin and/or glucagon secretion, or on the secretion of other gut hormones.

The other possibility is delayed clearance, and that involves that lipoprotein or hepatic lipase mechanisms.

Carroll: I tend to think that it is more related to degradation of cholesterol and its elimination from the system. This may be the primary effect rather than production.

Bodwell: Are there any differences between animals of different sex?

Carroll: We have used male rabbits for all of our studies and I really can't answer that question - somewhere along the line, one should look at it - male or female - but we haven't done it yet.

Martinez: I think one point with respect to the amino acids that should be recognized is that the plant proteins do contain a large proportion of their amino acids as glutamine and asparagine. This is a very real difference between the animal proteins and the plants.

Carroll: That is a very good point and the analytical data often don't make any distinction between asparagine and aspartic acid or glutamine and glutamic acid. I would be interested in having more information on that particular point.

Van Stratum: I'd like to comment a little bit on the rabbit model you used. We have done the same study and found the same relationships between casein, gluten, and soybean protein. One of the things that struck us was, like your study, we also had a low fat diet, and it turned out that the level of linoleic acid fat was about 1 percent of calories. We found that it presumably is a matter of essential fatty acid deficiency, because one of the characteristics of EFA deficiency is a very high serum cholesterol content; and as soon as small amounts of linoleic acid are added, something in the order of 3 to 5 percent of energy, there is a dramatic increase in growth, steep slope of serum cholesterol. Just working in that very sensitive area, any change in diet, be it small residual fatty acid or some change in amino acid which might be relevant, has a dramatic effect. We are analyzing the data now and this may explain a lot of the discrepancy between the animals, that is, the rabbits, and what we find in humans. The normal humans are relatively insensitive to a high level of soy proteins.

Carroll: Yes, we had some opportunity for informal discussion on this yesterday, and it has influenced my thinking
on the whole problem. However, as I showed on the one slide
where the diets contained 15% corn oil, there was still an appreciable difference between the plasma cholesterol levels on
the casein diet and the soy protein diet; but certainly the
low level of polyunsaturated fat may accentuate the difference
between the two, and this may be a factor in the human situation. I think that interactions between different dietary
components need to be considered in these experiments. However, I think that alterations in dietary protein offer a new
approach to treating hypercholesterolemia, and I personally
prefer the dietary approach for lowering plasma cholesterol
levels in humans to the drug approach. I think that physio-
logically it would be much better if it can be done by diet,
because this is something that has to be continued for a long
period of time. In my view, the dietary approach is preferable to having people taking drugs for twenty or thirty years.
I think this is a last-resort approach. I feel there is no
question that polyunsaturated fat can produce a definite lowering of plasma cholesterol levels; but it may be that one can
find dietary combinations that are more effective than changes
in any single dietary component by itself.

Iacono: I would like to push the point just made a little
further; in fact, I was coming up to make this same point. In
looking over your diets for the rabbits, they weren't on low
fat they were fat free, and in a true sense they are essential
fatty acid deficient. John Moore, from England, published
about ten years ago in the British Journal of Nutrition a paper where he fed diets very similar to the one you fed with
casein except he used 20% casein, I believe. He fed a high
casein, high sucrose diet. He showed that he could get a
great deal of atherosclerosis in these animals. By simply
adding 4% corn oil to his diet, he would eliminate the atherosclerosis. Now we have been feeding rabbits similar diets to
those of John Moore for years. When we use 20% casein
sucrose-starch diets with 4% corn oil our blood cholesterol
values are in the area of 68-80 milligrams percent. But what
surprised me in terms of what you have found is that in spite
of the fact that you had what I consider to be an essential
fatty acid deficient diet, that you are getting this difference in cholesterol between the casein diet and the soy isolate diet. A question that I would like to ask you is this.
Are there any essential fatty acids in the soy protein that
you use as your material for your diet and, if there are essential fatty acids in the soy preparation you are using,
might this not partially explain some of your results?

Carroll: Well, the information I have is that there is very little fat in the soy protein isolate that we have used, and I think that there was a slide shown earlier which indicated zero fat in soy protein isolate.

Struthers: I'd like to address this question since I did fat analysis on all of our soy isolates about three years ago. One of the things that we find is a considerable difference in the amount of fat extracted by acid hydrolysis and by ether extract. Using acid hydrolysis, you get some of the protein hydrolized as well as fat. Using ether extraction, only about 20-30% of the residual lipid is recovered. Chloroform-methanol methods yield the best results, and by chloroform-methanol you get around 1% fat in the soy isolates. Most of it is phospholipid and there are some essential fatty acids.

Rackis: I would like to take up the question I raised this morning. I think it would be very essential that the soy protein products be identified as completely as possible. In your rabbit diets, was all of the soy protein as soy protein isolate or was soybean concentrate or soybean meal used?

Carroll: In almost all experiments, it was an isolate. We did one experiment in which we compared a soy concentrate with a soy isolate, and the results were essentially the same. Thus, going from a concentrate to an isolate didn't seem to make any difference in our experience.

Rackis: There are many different soy protein products and they are manufactured under a wide range of conditions and, therefore, soy protein isolates would not necessarily have the same residual content; the unsaturated fatty content would be more or less. Unless the products are defined carefully, then it would be difficult for someone else to repeat the same experiment because some isolates can have up to 2-4% fat. The identification of the soy protein product becomes essential.

Carroll: All I have been saying with respect to the rabbit experiments is that we get the same results with an isolate obtained from Central Soya and an isolate obtained from Ralston Purina.

Harper: We did some work about fifteen years ago with different protein sources, mainly purified protein sources. We got fairly large differences in serum cholesterol concentrations with different protein in rats fed typical hypercholesterolemic diets with cholesterol and cholic acid added. When we extracted our protein sources most of the effects disappeared. There was tightly bound fat that was not extracted with ether particularly in the plant proteins; the wheat gulten was the most striking source. Some of the lipid material bound by the gluten was polyunsaturated fatty acids that was removed completely only after extraction with butanol.

Dr. Carroll, have you prepared your own products or are they just used as they are available commercially?

Carroll: Most of our protein preparations were obtained from other sources, and we haven't investigated their composition other than perhaps to see how much residual lipid could be extracted with chloroform-methanol. The alfalfa preparation (Table 1) was extracted in our laboratory with chloroform-methanol, followed by methanol, and gave plasma cholesterol levels in the same range as the other plant proteins. Some of the animal proteins we also extracted ourselves. For example, meats such as beef, pork, chicken and turkey were freeze-dried and then extracted with ether. This removed most of the lipid and reduced the cholesterol content almost to zero (Hamilton and Carroll, 1976).

Bressani: Two quick questions. First, I would like to know if the same effect could not be obtained from all of the legume foods like common beans, for example; and secondly, what about the rate of passage of vegetable proteins as compared to animal protein through the intestinal tract. I think vegetable proteins tend to move faster than the others. The faster rate of passage through the gastrointestinal tract may contribute to lower the levels of cholesterol.

Carroll: I'm not sure that I understood the question completely. You were asking about other legumes, and I gave results for pea protein and faba bean protein (Table 1). They gave the lowest plasma cholesterol levels, but contained about 5% fat, and I don't know whether this might have influenced the results (Huff et al, 1977a). It takes a lot of protein to feed six rabbits for a month, and to have extracted the residual lipid from this amount in the laboratory would require considerable time and effort, so we didn't attempt to do it for those particular proteins.

Bressani: Legume foods contain small amounts of fat and the fatty acid composition is highly unsaturated.

Carroll: I do not think so at that low level because, as I said, we have fed a diet of 5% corn oil in the casein and we still get the hypercholesterolemia at that level. Corn oil is about 50% linoleate, and safflower oil can run up around 70%;

but I think many of the oils run much higher than 50-60% linoleate. So, I think that 5% of the polyunsaturated fat may be
used. We haven't really worked it out, and perhaps we should
look at 8%, 10% and so on and see how much linoleate is required to produce an effect. But even 15% didn't seem to sup-
press the effect completely.

CRITIQUE OF METHODS FOR EVALUATION OF PROTEIN QUALITY

J. M. McLaughlan

Humans, like all other animals, require many nutrients including essential amino acids, vitamins, minerals and fatty
acids. Human requirements for the nine essential amino acids
have been estimated recently (World Health Organization, 1973)
and are similar to those proposed for other mammals. Except
in utero, amino acids are normally provided to the body
largely in the form of intact proteins which require digestion
to release their component amino acids and small peptides.
Proteins vary in their content of constituent amino acids. A
few proteins such as gelatin are completely devoid of one essential amino acid and when fed as the sole source of amino
acids will not sustain life. Long before the isolation and
characterization of individual amino acids, it was recognized
that gelatin was an inferior protein (Munro, 1964). It was
realized also that, in general, proteins of plant origin were
inferior to those of animals. This knowledge led to the concept of protein quality and methods concerned with the utilization of protein nitrogen were developed. Many people during
the past 75 years have been engaged in the sometimes frustrat-
ing problem of evaluating protein quality.

If research had followed a different course we might have
set requirements for each essential amino acid (the way we do
for vitamins) and the idea of "protein quality" might never
have arisen. In theory, we don't need a method for "protein
quality" - all we need are satisfactory methods for measuring
(available) amino acids and an accurate knowledge of human requirements for each essential amino acid. Block and Mitchell
(1946) pioneered this approach to evaluating food proteins,
calling the method chemical score. Unfortunately we are not
at the stage where we can rely completely on this approach.
Digestion of protein is not always complete and certain amino
acids, such as lysine and threonine, may not be fully available. In addition, estimates of human requirements for amino
acids also vary considerably. Consequently, we need a biolog-

ical assay for measuring the amount of the essential amino acid in greatest deficit relative to requirements - known simply
as "protein quality." In the near future we may be able to
dispense with animal bioassays when these problems with chemical score have been resolved.

Many biological methods have been proposed for evaluating
protein quality for humans and several are in use today. Various animals (human, pig, rat, mouse, tetrahymena, etc.) have
been utilized for this purpose but only methods dealing with
young growing rats will be considered in this review. Young
and Scrimshaw (1974) reviewed the evidence justifying the rat
assay and concluded that the data supported the use of the
growing rat for estimating protein quality for humans.

Thomas (1909) provided us with the first true method for
assessing differences in protein quality - the biological value (BV) method. Mitchell (1924) standardized the procedure
and applied it to many foods. It is the classical method for
protein evaluation and although laborious, it is still carried
out in some laboratories. The customary equation given for BV
is disarmingly simple.

$$BV = \frac{\text{Retained N}}{\text{Absorbed N}} \times 100$$

A more meaningful equation is:

$$BV = \frac{\text{N intake-(fecal N-endog. N)-(urine N-endog. N)}}{\text{N intake-(fecal N-endogenous N)}} \times 100$$

Biological value is a nitrogen balance technique which
distinguishes between ingested protein and absorbed protein.
Endogenous nitrogen in urine and feces are estimated when animals are fed a protein-free diet. The procedure has been applied to several species of animal, including humans and rats.
The method is laborious, subject to several possible sources
of error, and is not suitable for the routine assessment of
protein quality.

Anyone who has carried out protein evaluation with young
rats knows that rat growth provides one of the simplest biological assays for this purpose. Osborne et al (1919) increased the dependability of the rat growth method by expressing the weight gain and protein consumed as a ratio - hence
the name "protein efficiency ratio" or PER.

$$PER = \frac{\text{Weight gain}}{\text{Protein consumed}}$$

PER values for a protein increase as the protein content

of the diet is raised up to a maximum value (usually about 9–10% protein) and then decrease as the protein level is raised further. Osborne and co-workers (1919) carried out PER determinations at a few protein levels and accepted the highest value as the PER value for the protein. PER has come into widespread use, but only one arbitrary level of protein (9% or 10%) is commonly fed. Standardization of PER at one level penalizes both high quality and low quality proteins (i.e., whole egg and wheat flour may have highest PER values at 8% and 12% protein respectively). Although PER is the official method in the United States and Canada, it has serious shortcomings which are generally well-known and will be discussed later.

Miller and Bender (1953) proposed net protein utilization (NPU) which assessed the efficiency of nitrogen utilization of the test protein. The carcass nitrogen of the test group and a group fed a non-protein diet is measured.

$$NPU = \frac{\text{body N of test group} - \text{body N of group fed non-protein diet}}{\text{N consumed by test group}}$$

Determination of body nitrogen is a messy business, and several groups of workers (Bender and Miller, 1953; Dreyer, 1957; Henry and Toothill, 1962; Hegsted et al, 1968) have shown that body nitrogen can be calculated accurately from the more simply determined body water. Others have shown that in short-term experiments, such as 10–14 days, body weight is an accurate reflection of body nitrogen (Campbell, 1963; Hegsted et al, 1968). NPU is widely used, particularly in Europe.

Bender and Doell (1957) published a simpler version of NPU called net protein ratio or NPR. Body weight rather than body nitrogen was measured.

$$NPR = \frac{\text{body weight of test group} - \text{body weight of group fed non-protein diet}}{\text{protein consumed}}$$

The similarity between NPR and NPU is obvious although the magnitude of the actual values are on different scales due to the parameters measured.

The more usual equation for NPR is:

$$NPR = \frac{\text{gain in weight of test group} + \text{loss in weight of groups fed non-protein diet}}{\text{protein consumed}}$$

This latter equation is the same as PER except for the inclu-

sion of the weight loss of the non-protein group. Both NPR
and NPU are two-dose assays (i.e., the test protein and a zero
protein level).

Hegsted and Chang (1965a, 1965b) proposed the slope ratio
assay which is a multi-dose procedure which includes a non-
protein diet and three or more dietary levels of protein.
Only levels falling on the linear portion of the response
curve are used in computing the slope. Lactalbumin was chosen
as a reference protein and the slope of the test protein was
expressed as a percent of the slope for lactalbumin.

All of the methods discussed measure either directly or
indirectly the efficiency of utilization of the food protein
for deposition of body nitrogen. Similarities and interrela-
tionships among PER, NPU and NPR and slope ratio are illustrated in Figure 1. The parameters measured are food eaten,
from which nitrogen or protein intakes are calculated, and
change in body weight (or body water or body N). Hegsted et
al (1968) carried out slope ratio assays for several foods
measuring body N, body water and body weight. Results for
these protein sources calculated from slope ratio assays using
the three parameters, body N, body water and body weight
agreed closely, but the error was least when body weight was
the parameter. The left hand frame shows six doses in a slope
ratio assay; the top dose would be ignored. The slope of the
response line is a/b. Factors involved in NPR and NPU are
shown in the right hand frame. NPR is the weight gain of the
test group plus the weight loss of the non-protein group over
the protein consumed. It is obvious that NPR is also an expression of the slope of the response line, and NPR and NPU
are simply two-dose versions of the slope ratio assay.

PER is the weight gain of the test group/protein consumed
and is not the slope of the response line. The PER value at
zero weight gain is zero. Therefore, PER values determined at
different protein levels are not proportional. Numerically a
PER of 1.0 is 100 times more than a PER of 0.01. With high
quality proteins supporting a large gain in weight, PER values
(relative to high quality proteins such as lactalbumin) are
only slightly lower than NPR values.

Protein quality measures for 23 protein sources are given
in Table 1. The value for casein was set at 100 for each type
of assay. The first five samples (FM16, SM13, SM7, SM3 and
soya plus methionine) had numerically similar values by PER,
NPU and NPR. For poor quality proteins shown at the bottom of
the table, PER yielded much lower values than the other methods. Since NPU and NPR are essentially the same assay, agreement between these methods was excellent.

Table 2 shows PER, relative protein value (RPV) and NPR
data for several samples. RPV is the same as the slope ratio

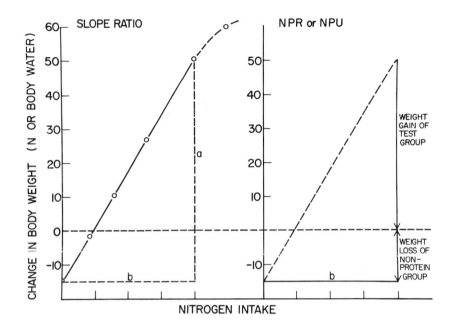

FIGURE 1. The bases of slope-ratio, NPR and NPU assays for protein quality. In the NPU assay body N rather than body weight is used. (From J. M. McLaughlan in Newer Methods of Nutritional Biochemistry (A. A. Albanese, ed.) Academic Press, New York, pp. 33-64.)

(SR) assay except that the data for the non-protein group is omitted when calculating the slope of the response lines (Samonds and Hegsted, 1977). Except for high quality protein, PER gave lower relative values than the other two methods. Agreement between RPV and NPR was excellent except for casein, white flour and wheat gluten. The reasons for these differences are discussed later.

Statements have been made about the relationship between PER and other methods, and this relationship has been misused in discussions concerning regulations for protein quality. The argument goes like this: "If soya protein has a PER 70% of casein and casein has a PER of only 70% that of lactalbumin, then the soya really has a PER 49% of the value for lactalbumin. Everyone knows that even poor quality proteins such as white flour have NPU's of approximately 50." Clearly this is not a valid comparison. The trouble with PER is that values are not proportional and it is obvious that PER is not a

Table 1.   Relative PER, NPU and NPR values for 23 protein
           sources (casein = 100)

Protein Source	PER	NPU	NPR
Casein	100	100	100
FM 16	109	100	103
SM 13	100	105	109
SM 7	99	103	108
SM 3	98	109	110
Soya + Methionine	97	106	103
SM 8	84	100	96
SB 5	83	83	91
Soya	79	85	86
SB 1	79	84	91
CM 49	76	94	94
Yeast	63	78	79
SF 2	63	85	82
SF 3	62	91	88
CM 19	60	87	89
Linseed cake	59	95	91
GN 12	48	66	69
SB 3	47	67	71
GN 18	43	60	61
SF 1	38	81	80
GN 11	32	55	50
CM 13	25	71	74
Wheat gluten	19	64	71

FM = fish meal, SM = dried skim milk, SB = soya, CM =
cottonseed meal, GN = groundnut and SF = sunflower seed meal.
   Values expressed as a percentage of value obtained for casein.
   From Henry and Toothill (1962).

suitable measure of protein quality – at least for regulatory
purposes.
   There are probably only two methods worthy of serious consideration as a replacement for PER. These are the slope assay (in its two forms, SR and RPV) and the NPU-NPR type of assay. If all protein sources yielded valid slope ratio assays
(i.e., straight line dose-response curves meeting at the zero
protein level on the y-axis), only two doses would be needed.
Either NPU or NPR would be suitable. Experience with the SR
assay has shown that certain types of proteins yield invalid

Table 2. *Protein quality of 15 protein sources by three methods*

Protein Source	Relative PER	RVP	Relative NPR
Lactalbumin	100	100	100
Macaroni + cheese + threonine	97	100	102
Macaroni + cheese	67	85	83
Casein	72	92[1]	82
Meat	68	74	77
Soy #1	70	67	75
Soy #2	62	72	72
Oat flour	55	65	67
White flour	23	32	46
Wheat gluten	5	20	37

[1] *The value for casein was 80 in the SR assay.*
Value expressed as a percentage of value obtained for lactalbumin.

assays due to excessive downward curvature. Specifically, these are lysine-deficient proteins such as wheat gluten (McLaughlan and Campbell, 1969; Yanez and McLaughlan, 1970). This is a serious problem for both SR and NPU-NPR. Statistical treatment of such data shows the SR assay to be invalid. In the SR and NPR assays, the values fall when the protein is tested at increasingly higher levels. The most extreme case observed is illustrated in Figure 2. The sample was a toasted high-protein breakfast cereal in which lysine was almost certainly damaged. Only the lower level of the casein response curve is shown. NPR values when the product was tested at 5%, 10% and 15% levels of dietary protein were 2.75, 1.56 and 1.25, respectively. The SR assay yielded a non-valid assay due to excessive downward curvature. To circumvent this situ-

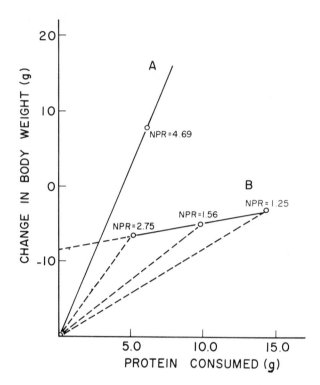

FIGURE 2. NPR data for casein and a breakfast cereal at different protein intakes. The NPR for casein (response line A) was 4.69 when fed at the 5% level of dietary protein. NPR values for the breakfast cereal (response line B) were 2.75, 1.56 and 1.25 when tested at 5%, 10% and 15% levels of dietary protein respectively. (From McLaughlan, 1974, in Nutrients in Processed Foods-Proteins (P. L. White and D. C. Fletcher, eds.) Publishing Sciences Group, Inc., Acton, Massachusetts.)

ation, it was suggested that the zero protein group be omitted when calculating the slope (McLaughlan, 1974; Samonds and Hegsted, 1977). Samonds and Hegsted (1977) called this version of the slope assay relative protein value or RPV.

However, greater experience with RPV leads to the conclusion that SR is a more suitable method than RPV. The chief problem with RPV is parallelism (McLaughlan and Keith, 1975; McLaughlan and Keith, 1977). Parallelism exists when the test protein response curve cuts the y-axis at a point more negative than for the reference protein. The RVP assay yields high values for such samples. This situation occurs frequently with mixtures of protein limiting in threonine (McLaughlan and Keith, 1977; McLaughlan, 1977) and with casein. In Table 2, the RPV value for casein was 92 whereas the SR and NPR values were 80 and 82, respectively. In the RPV method the intercepts on the y-axis were -13.7 and -6.7 for casein and lactalbumin, respectively. Parallelism such as this yields erroneously high values. Hackler (1977) reported marked differences in slopes and apparent protein quality between SR and RPV assays. Values for peanut meal were 0.51 and 0.71 by SR and RPV, respectively. Tuna fish had a value of 0.82 by SR and 1.07 by RPV. There are other clear-cut cases of parallelism in the recent literature (Chavez and Pellett, 1976; Chang and Field, 1977) if the appropriate data are plotted. Parallelism is largely eliminated when the data for the protein-free group is used in calculating the slope (i.e., SR method).

McLaughlan and Keith (1977) concluded that the RPV underestimated the protein quality of lysine-deficient proteins. Food mixtures such as rice-casein, peanut-sesame-fish and soya protein with added methionine were formulated to be approximately co-limiting in lysine and threonine. At low protein levels supplementation with threonine increased growth significantly but did not increase growth at the high protein level - presumably lysine was limiting at the 8-9% protein level. Increased growth at the low protein level and none at the high level due to threonine supplementation resulted in decreased slopes in the RPV assay.

In similar studies with whole wheat flour (McLaughlan, 1977,) lysine and/or threonine were used. The slopes and intercepts (y-axis) for the response lines are given in Table 3. The most striking observation is the wide divergence on the y-axis of the intercepts with the RPV method. The weight loss of the group of rats fed the non-protein diet was -9.9, and the intercepts of the response lines were close to that value in the SR assay.

The slopes of the response line for the basal diet in the RPV and SR methods were 1.79 and 2.07, respectively. Inclusion of the zero protein group in calculating the slope in the SR assay increased the slope. Addition of lysine markedly increased the slope in the RPV method. Lysine supplementation did not increase growth at the 4% protein level. Consequently, the response line rotated about the 4% level and the intercept decreased to -12.17. Supplementation with threonine increased growth at the low protein level (significant at 1%), but did not increase growth at the 10% protein level; therefore, the slope in the RPV assay was decreased and the intercept was -5.44. When the non-protein group was used in calculating the slope in the SR method threonine addition had little or no effect on the slope. Addition of threonine made lysine the sole limiting amino acid, and the resultant horizon-

Table 3.   Slopes and intercepts for whole wheat flour samples
            in RPV and SR assays

Amino acid(s) added	RPV[1]		SR[2]	
	Slope	Intercept[3]	Slope	Intercept[3]
Glutamic acid	1.79	− 8.18	2.07	− 9.76
Lysine	3.06	−12.17	2.92	−11.26
Threonine	1.46	− 5.44	2.13	− 9.17
Lysine & Threonine	3.32	− 6.29	3.61	− 9.17

[1]The data from rats fed the non-protein diet were not
included in calculating the slope.
[2]The data from rats fed the non-protein diet were included in calculating the slope.
[3]Intercept on y-axis of the response line. The protein
levels tested were approximately 0%, 4%, 7% and 10% with eight
rats per protein level.
From McLaughlan, 1977.

tal type of slope in the RPV assay is typical of the kind
found with severely lysine-deficient proteins (Yanez and
McLaughlan, 1970).
    The apparent protein quality by PER, relative NPR, SR and
RPV methods for these samples are given in Table 4. The values for the basal diet (glutamic acid) were arbitrarily set at
100. Lysine supplementation increased PER and RPV markedly;
relative NPR and SR gave almost identical values. Threonine
supplementation had no effect on PER, NPR or SR, but decreased
the value in the RPV assay. Relative NPR, SR and RPV yielded
similar values with lysine and threonine supplementation.
    It is obvious that the protein quality of whole wheat
flour was not decreased by supplementation with threonine, yet
the RPV assay indicated this to be so. These data support the
previous finding that the RPV method underestimates the quality of lysine-deficient proteins that give the horizontal type
of response. One of the interesting aspects of this study was
the close agreement between relative NPR and SR for lysine-
deficient proteins. The good agreement is due in part to
testing proteins at comparable dietary levels. Relative NPR

Table 4. PER, relative NPR, SR and RPV values for whole wheat flour supplemented with amino acids

Amino acid(s) added	PER	Relative NPR	SR	RPV
Glutamic acid	0.65 ± 0.08[1]	100	100	100
Lysine	1.64 ± 0.15	142	141	171
Threonine	0.75 ± 0.07	107	103	82
Lysine & Threonine	2.83 ± 0.08	180	174	185

[1]*SEM.*
From McLaughlan, 1977.

and SR are in reasonably good agreement for lysine-deficient proteins when the high dosage level is 8-10% protein, but the slope decreases in the SR assay when the protein is raised. However, few cereals can be fed above 10% protein in the diet wihout resorting to some method for concentrating the protein. In any case, Jansen (1978) has questioned the relevance of these low SR and RPV values in human nutrition. It is fair to conclude that the severely lysine-deficient proteins are a problem with all the assays being discussed.

There is one other criticism of the RPV assay, and that is the flexibility in choosing dietary protein levels (McLaughlan, 1977b). The purpose of the flexibility is to permit the analyst to choose the widest possible linear range of dosage levels (Samonds and Hegsted, 1977). For instance, white flour and wheat gluten (Table 2) were tested up to 13% and 27% dietary protein levels, respectively. Usually 3%, 5.5% and 8% levels for lactalbumin are chosen, resulting in slight but non-significant curvature for the reference protein. By judicious selection of dosage levels for test protein and standard, the apparent value of the sample can be manipulated. For instance, choosing levels of 3%, 6% and 9% for the standard may give a lower slope than if the levels 2.5%, 5.0% and 7.5% protein were chosen. Similarly, the slope of the test protein can be altered. Presumably, this juggling of protein levels would be legal unless dietary levels are specified. This criticism is not valid with the SR assay because the zero-protein dose firmly fixes one end of the response curve.

Another point to consider with the RPV method is that the middle dose of a three-dose assay has almost no effect on the slope - yet it is the slope that is the measure of protein quality. Therefore, 1/3 of the data is wasted; its only function is to indicate the extent of curvature. If there is curvature, what is the course of action? If curvature is due to a deficiency of lysine or some other amino acid, then presumably repeating the assay would simply produce another similar assay with curvature. Curvature results from joining points that reflect different entities. For lysine-deficient proteins, quality decreases as the protein content of the diet is raised and presumably is due to a difference in lysine requirements for maintenance and growth (McLaughlan and Campbell, 1969).

It appears that the multi-dose slope assay has been oversold as the only satisfactory method for evaluating protein quality. Its weakness has been illustrated - perhaps even exaggerated the problems. Nevertheless, the slope assay is not that much better than NPR to justify the extra expense. The Health Protection Branch in Ottawa is probably going to choose NPR to replace PER as the official method.

This version of NPR is relative NPR (PAG, 1975). The test protein is evaluated at 8% in the diet with casein plus methionine as the reference protein. Recently a collaborative assay comparing PER and relative NPR was conducted (McLaughlan et al, 1978). Collaborators were given the test proteins and approximate protein content. Each collaborator prepared his own diets conducting the appropriate nitrogen analyses. Apparent differences in the protein content of diets was larger than expected, and it seems likely that part of the error in the collaborative assay was due to differences in estimating the protein content of diets. NPR was superior to PER and only the data for NPR are shown in Table 5. Agreement among collaborators was reasonably good considering that ready-made diets were not provided. The value of 32 for wheat gluten is considerably higher than the 20-25 usually obtained with the RPV method. As pointed out in this review the RPV method may underestimate protein quality of lysine-deficient foods. The important question is whether or not the RPV method is more predictive for humans than the NPR estimate. In a recent review Jansen (1978) has stated "The data suggest that even for lysine-deficient white flour, the NPR does not overestimate its value for the infant and, in fact, NPR is a better predictor for the human infant than is the RPV."

Samonds and Hegsted (1977) believe that the multi-dose slope assay is more appropriate than NPR because they claim it has more of the characteristics of a good bioassay. That may be true - in theory. However, protein quality depends princi-

Table 5.   Relative NPR Value

Collaborator	Lact- albumin	Egg white	Wheat gluten	Soya isolate	Soya + wheat gluten
1	92	103	34	66	73
2	96	105	38	64	66
3	88	97	33	62	66
4	94	101	25	59	61
5	77	94	32	47	61
6	77	84	33	58	60
Mean	87.3	97.3	32.5	59.3	64.5

pally on the content of the limiting amino acid and the relative requirements of that amino acid for maintenance and growth. Because the limiting amino acid differs with various proteins and sometimes with different dietary levels of the same protein, methods for the evaluation of protein are inherently non-specific. It is naive to expect to develop a fully satisfactory assay for protein quality because protein quality is not a fixed entity.

REFERENCES

Bender, A. E., and Miller, D. S. (1953). *Biochem. J.* 53, 7-8.

Bender, A. E., and Doell, B. H. (1957). *Brit. J. Nutr.* 11, 140-148.

Block, R. J., and Mitchell, H. H. (1946). *Nutrition Abstr. Revs.* 16, 249-278.

Campbell, J. A. (1961). Methodology of protein evaluation. *Nutrition Doc. R. 10/Add. 37, WHO/FAO/UNICEF-PAG*.

Chang, Y., and Field, R. (1977). *J. Nutr.* 107, 1947-1953.

Chavez, J. F., and Pellett, P. L. (1976). *J. Nutr.* 106, 792-801.

Hackler, L. R. (1977). *Cereal Chem.* 54, 984-995.

Hegsted, D. M., and Chang, Y. (1965). *J. Nutr.* 85, 159-168.

Hegsted, D. M.,; Neff, R.; and Worcester, J. *J. Agric. Food Chem.* 16, 190-195.

Henry, K. M., and Toothill, J. (1962). *Brit. J. Nutr.* 16, 125-133.

Jansen, G. R. (1978) personal communication.

McLaughlan, J. M., and Campbell, J. A. (1969). In *Methodology of Protein Evaluation in Mammalian Protein Metabolism* (H. N. Munro, ed.). Academic Press, New York, pp. 391-422.

McLaughlan, J. M. (1974). Evaluation of standard rat assays in *Nutrients in Processed Foods - Proteins* (Philip L. White and Dean C. Fletcher, eds.). Publishing Sciences Group, Inc., Acton, Massachusetts, pp. 69-76.

McLaughlan, J. M. (1977). *Nutr. Rep. Int.* 16, 439-445.

McLaughlan, J. M. (1978). The problem of curvature in slope assays in "Protein Nutritional Quality Improvement." (Mendel Friedman, ed.). Marcel Dekker, Inc., (in press).

McLaughlan, J. M., and Keith, M. O. (1975). Biossays for protein quality in *Protein Nutritional Quality of Foods and Feeds,* Vol. 1 (Mendel Friedman, ed.). Marcel Dekker, Inc., New York, pp. 79-85.

McLaughlan, J. M., and Keith, M. O. (1977). *J.A.O.A.C.* 60, 1291-1295.

McLaughlan, J. M.; Anderson, G. H.; Hackler, L. R.; Hill, D. C.; Jansen, G. R.; Keith, M.O.; and Sosulski, F. W. (unpublished findings).

Miller, D. S., and Bender, A. E. (1955). *Brit. J. Nutr.* 9, 382-388.

Mitchell, H. H. (1924). *J. Biol. Chem.* 58, 905-922.

Munro, H. N. (1964). Historical Introduction. The origin and growth of our present concepts of protein metabolism in Mammalian Protein Metabolism (H. N. Munro and J. B. Allison, eds.). Academic Press, New York, pp. 1-29.

Osborne, T. B.; Mendel, L. B.; and Ferry, E. L. (1919). *J. Biol. Chem.* 37, 223-229.

Protein Advisory Group (PAG). Statement No. 28 (June 1975). Vol. 5. No. 2, Protein-Calorie Advisory Group, United Nations System, pp. 22-48.

Samonds, K. W., and Hegsted, D. M. (1977). Animal bioassay: A critical evaluation with specific reference to assessing nutritive value for the human in Evaluation of Proteins for Humans (C. E. Bodwell, ed.). The Avis Publishing Company, Inc., Westport, Connecticut.

Thomas, K. (1909). *Arch. Anat. Physiol. Abt.* p. 219.

World Health Organization, Tech. Rep. Ser. (1973), No. 522. Energy and Protein Requirements Report of a Joint FAO/WHO Ad Hoc Expert Committee, Geneva.

Yanez, E., and McLaughlan, J. M. (1970). *Can. J. Physiol. Pharmacol.* 48, 188-192.

Young, V. R., and Scrimshaw, N. S. (1974). Relation of animal to human assays of protein quality in Nutrients in Processed Foods-Proteins (P. L. White and D. C. Fletcher, eds.). Publishing Sciences Group, Inc., Acton, Massachusetts., pp. 85-92.

DISCUSSION

Harper: Why can't we go right now to amino acid score and abandon the rat assay?

McLaughlan: We haven't as yet have tried and tested methods for measuring the digestibility of proteins and availability of amino acids.

Samonds: About 8 years ago, Mark Hegsted wrote a letter to D. J. Finney, the "architect" of the slope-ratio procedure, explaining the problems we were finding with lines having no common intercept. In very elegant British and statistical language, Finney responded to the effect that "you can't measure potency." Sometimes I wish I had heeded that advice and had gone to work on something simple - like flatulence factor! Since protein is composed of amino acids with different abilities of recycling, we are forced into the situation described by Dr. McLaughlan where the dose-response lines deviate from the ideal situaton required for a valid assay...and I agree with Dr. McLaughlan by paraphrasing George Orwell, "All assays are invalid, but some assays are more invalid than others." However, I'm not any happier with Dr. McLaughlan's solution to the problems than I am with the problems themselves. Forcing a regression line through an intercept other than the intercept indicated by the data for a given protein, as he proposes in the Relative NPR, does several things. First, it tends to increase the variability about the line because that line is not the Least-Squares Estimate of the best fit. An increase in the standard error of the slope estimate would reduce our confidence in a potency estimate and reduce our ability to discriminate between proteins. Next it may lead to significant curvilinearity if one uses the appropriate statistical tests to identify curvature...which NPR does not. Third ... although this procedure might tend to decrease variability (in contrast to my first point) by extending the range of the data used to calculate the regression equation, this only leads to a false sense of security about one's data. Finally, and perhaps most important, this method overemphasizes the importance of the response at very low intakes where the limiting amino

acid may not be the one which is limiting in the range of intakes generally consumed; we are reaching a conclusion about a
protein's ability to promote growth based upon the results of
animals which are losing weight.

I have proposed earlier that we consider two measures of
protein quality - a potency for maintenance calculated as
Dr. Scrimshaw does his Nitrogen Balance data, and a potency
for growth based upon the slopes of regression lines of animals or children accumulating body protein. A few questions
have arisen, however. How do we use this approach? Is maintenance for children the same as maintenance for adults? Are
we making a complicated mess out of a simple mess?

I think our methodological squabbles may be an exercise in
futility until we show some relevancy of rat bioassays to human protein utilization. But I don't think that proposals to
junk the animal assays are realistic in light of the high cost
and effort of human experimentation nor are our hydrolysis
procedures or estimates of human amino acid requirements reliable enough at present to permit the sole use of Amino Acid
Score.

Pellet: I am glad that Dr. McLaughlan made the point that
had history been different we would be concerned with amino
acid requirements rather than protein quality. I disagree
with him, however, that the time is not yet for switching to
the more fundamental consideration of amino acids rather than
protein quality as such.

Protein quality determination has too many variables in
both its determination and interpretation. Even when protein
is assayed using the best available procedures, ancillary
studies, both biological and chemical, are necessary to assess
fully the potential value of a protein in a real-life situation. Health, nutritional status, age, and physiological
state of the individual consuming the protein, together with
the complete dietary composition, including other proteins,
and the total energy value, can all combine and interact to
affect the final value of the protein to the consumer.

Because of these complex interrelations, I am sure that we
have been on the wrong track in attempting to use bioassays
(both human and animal) as our reference criteria for the
quality of protein. Since every other dietary component consumed with a protein source and the individual circumstances
of the consumer affect the quality as measured, an absolute
value for protein quality true under all circumstances is an
unattainable goal, and that searching for such a technique is
regressive rather than progressive. I am fully convinced that
we should concentrate on the definable characteristics of the
protein, i.e., total nitrogen, total and available amino acid

composition (perhaps eventually even sequence), and digestibility (defined in terms of secondary and tertiary protein structure), as our criteria of quality. These values, while not absolutely constant, are more so than are biological responses. A protein very low in lysine remains very low in lysine whether or not a consuming animal is able to adapt and reutilize lysine more efficiently on a short-term basis. This is similar to considerations of the iron content of a dietary which remains constant irrespective of the degree of absorption and utilization which may occur.

Criteria solely based on the protein must, of course, be modified in practice by biological and dietary considerations. Nevertheless, if we continue to define the quality as measured by PER or even RPV as 'correct' and chemical determinations as 'incorrect' if they do not agree, the problems will never be resolved.

We have gotten ourselves into a corner with PER and its role in regulation. Prediction of quality from amino acid data is fully ready for use and, in conjunction with digestibility, could be more fundamantal and more accurate than continuing to worry about the interpretation of bioassays. Data for lysine, total sulfur amino acids and perhaps threonine are all that are necessary for the vast majority of food proteins and, when expressed as percentages of requirements, have the tremendous advantage of being *additive* for mixed complementary dietaries, an attribute that is impossible for bioassays, no matter how accurate, to emulate.

HOW SHOULD PROTEIN QUALITY FOR HUMANS BE DETERMINED

Consideration by Panel
D. T. Hopkins, Chairman

Vegetable proteins can assume many forms and perform many functions in food systems as has been discussed in earlier papers at this conference by Waggle and Martinez. Functionality, of course, refers to attributes of proteins such as color, flavor, texture, solubility, gel forming ability, and water binding capacity. If vegetable proteins can compete with other ingredients in terms of functionality and cost, then food manufacturers will be able to process foods containing vegetable protein that are acceptable to the consumer. This seems to be the way that the consumption of vegetable protein by people will increase.

The protein quality of one of the most important vegetable proteins - soy protein - has been measured in feeding studies with human adults, children, and infants. These studies, which have been reviewed in earlier sections of this book, have demonstrated that soy protein is a highly nutritious source of protein, and has an excellent quality in terms of meeting the amino acid requirements of humans (Fomon and Ziegler, Torun, Scrimshaw and Young). Jansen discussed the importance of protein quality in human nutrition and concluded that high quality is indeed a worthwhile goal for proteins to be fed to people. On the other hand, Harper pointed out that at least one of the primary purposes of feeding proteins to people is to meet their lysine and sulfur amino acid requirements. Thus, one might reason that protein quality basically is a function of the level of lysine and sulfur amino acids in the protein. Furthermore, McLaughlan has pointed out that there is considerable controversy on methods to measure protein quality.

Most methods used to measure protein quality are based upon some type of feeding study using the growing laboratory rat. In addition to the question of what techniques and procedures should be used to measure protein quality, we must

299

Table 1. The Protein Efficiency Ratio (PER) of Some Proteins

Dried Whole Egg	3.1
Dried Egg White	2.8
Lyophilized Beef	2.7
Casein	2.5
Oat Flour	1.9
Isolated Soy Protein	1.8
Peanut Meal	1.7
Potato Flour	1.7
Durum Wheat Flour	0.7
White Wheat Flour	0.5

also ask if laboratory animals such as the rat are good models for measuring protein quality. In Table 1 are shown PER values of a variety of animal proteins and vegetable proteins measured in our laboratory. As can be seen from this list of PER values, vegetable proteins usually have PER values of less than 2.5, whereas animal proteins have PER values higher than 2.5. Does this mean that vegetable proteins are inherently poorer quality than animal proteins? Several recent human feeding studies indicate that this is not the case.

On the following pages is the panel discussion on procedures that should be used to measure the quality of proteins for human feeding. In this discussion the panel examines the suitability of animal studies for measuring the quality of protein to be fed to humans as well as the possibility of developing appropriate *in vitro* assays.

In the discussion that follows, a number of abbreviations are used to signify different methods of measuring protein quality. A brief glossary follows for the reader who is not readily informed as to the meaning of all the terms.

PER - Protein Efficiency Ratio, NAS-NRC (1963)
BV - Biological Value, NAS-NRC (1963)
RPV - Relative Protein Value, Samonds and Hegsted
RNU - Relative Nitrogen Utilization, McLaughlan (1964)
NPU - Net Protein Utilization, NAS-NRC (1963)

REFERENCES

Fomon, S. J. and Ziegler, E. E. (1978). Soy Protein Isolates in Infant Feeding. This volume.

Harper, A. E. (1978). Human Requirements for Lysine and Sulfur-Containing Amino Acids. This volume.

Jansen, G. R. (1978). The Importance of Protein Quality in Human Nutrition. This volume.

Martinez, W. H. (1978). The Importance of Functionality of Vegetable Protein in Foods. This volume.

McLaughlan, J. M. (1976). J.A.O.A.C. 59: 42–45.

McLaughlan, J. M. (1978). Critique of Methods for Evaluation of Protein Quality. This volume.

NAS/NRC (1963). Evaluation of Protein Quality. Publication 1100, Washington, DC.

Samonds, K. W. and Hegsted, D. M. (1977). Evaluation of Proteins For Humans (Bodwell, C. E., ed). The Avis Publishing Company, Inc., Westport, CT.

Scrimshaw, N. S. and Young, V. R. (1978). Soy Protein in Adult Human Nutrition: A Review With New Data. This volume.

Torun, B. J. (1978). Nutritional Quality of Soybean Protein Isolates: Studies in Children of Preschool Age. This volume.

Waggle, D. H. (1978). Types of Soy Protein Products. This volume.

NEEDS AND CONCERNS BY FEDERAL REGULATORY AGENCIES
ON MEASURING PROTEIN QUALITY

John E. Vanderveen

The need for federal regulatory agencies to set quality
standards and measure protein quality in the interest of the
consumer is obvious. The questions that need to be answered
are what are the standards which are needed to meet consumers'
needs and expectations; what are the methodologies which can
be used to assure that the standards are being met; and what
are the risks that present knowledge may be inadequate to accurately assess the consumers' requirements?

Clearly, the most critical needs among our population are
those of the infant and young growing child. We must not only
be concerned that the amino acid needs of this group be adequate to support optimum growth and mental development; we
must also be sure that their needs are met for maintaining a
viable immune response for protection against viral, bacterial
and other microbiological diseases. In addition, we must in-
sure their ability to ward off the effects of a variety of
toxic substances with which the individual routinely comes in
contact in our modern society. An adequate supply of quality
protein frequently provides protection against these substances.

The Food and Drug Administration (FDA) finds a lack of
consensus of scientific opinion on this issue of protein quality measurement which can be interpreted as a need for additional scientific research. It is essential that an adequate
scientific data base be established before any public rule
making can be initiated to change existing regulations. This
conference has represented a giant step in the direction of
creating that data base; however, it is obvious that much more
work remains to be done. In my opinion, the scientific community has to make a meaningful effort to resolve this deficiency in knowledge and to agree on interpretations so that
the process of public rule making can be initiated. The continued absence of needed data and consensus of scientific

opinion will likely result in legal or political actions to
dictate the future use of protein products in the marketplace.

In addressing the question of which methodology is appropriate for measuring the quality of proteins, regulatory agencies must make certain that the selection criteria include
consideration of the needs of all segments of our society,
with particular emphasis on infants and young children. The
current approved methodology, the protein efficiency ratio
(PER), has been criticized for not accurately assessing the
biological value of proteins with respect to human requirements. The data derived from human studies presented at this
meeting tend to support these claims.

The preamble to the final regulation on vegetable protein
products will respond to this situation this way:

"The Commissioner recognizes the limitations of the PER
method. The applicability to human nutrition of the PER assay
and other rat growth assays commonly used to estimate the protein quality of foods has been questioned. It is recognized
that there is a lack of adequate data regarding the physiological amino acid and nitrogen requirements of both man and animals and the factors which may affect these requirements.
Scientific information on the protein quality of foods in humans is of a highly variable nature because of methodological
uncertainties and the lack of standardization between experiments. The development and/or improvement of methods to be
used to more adequately evaluate the protein quality of foods
in human nutrition depends greatly on the availability of more
precise and extensive data on the amino acid requirements of
humans and animals and on animal research techniques which can
be used to compare with human data. Since many of the protein
foods included in this regulation will be consumed by children, it is appropriate to use an assay that uses a growing
animal for test purposes.

"The PER assay is the only official method currently
available for determining protein quality. There are other
methods being used to measure protein quality, but the merits
of these methods have not yet been established through the
collaborative studies which are necessary for them to be made
official. The Commissioner therefore concludes that for the
purposes of this regulation, it is appropriate to use the PER
assay as the basis for determining protein quality."

The PER assay is also used for all other protein quality
requirements in FDA regulations (Vanderveen, et al, 1977). Up
to this time there is no indication that PER has in any way
compromised the safety of the consumer, and if there is an
error, it is in favor of consumer's health. This is not to
say that PER should not be changed or replaced. It simply
means that in the absence of improved methodology which safe-

guards human needs, continued use of PER is appropriate from a regulatory point of view. As indicated above, regulatory agencies are concerned about the deficiencies of the PER measurement and are conducting significant research to develop new methodology; however, resources are limited and Government cannot and should not conduct all needed research. Industry has both a responsibility to the public and a vital interest in the development of new methodology. It is imperative that industry continue to provide a significant share of the cost of, and effort in, this research.

In considering guidelines for this research, it is obvious that any new methodology must accurately measure protein quality with respect to human requirements. In addition, the following criteria are important from a regulatory point of view:
1) It must be possible to accomplish the assay in a reasonable period of time (no longer than two weeks);
2) the assay not require the use of human subjects;
3) the assay must include an assessment of bioavailability; and
4) The procedure should be endorsed by the AOAC or other appropriate body relative to accuracy and repeatability. The latter condition is not absolutely essential; however, such an endorsement would likely speed up the process of acceptance.

Finally, I would like to address the question of risks that present knowledge may be inadequate to assess the consumer's requirements. Despite extensive research which has already been reported, it seems clear from discussions at this meeting that more research is required to define precisely human requirements for amino acids. Most research accomplished to date has been conducted on highly selected individuals who were free of obvious disease or stress. However, we must be sure that the needs of the entire population are considered in regulating the food supply. Consequently, allowance for deviations in requirements must be included in regulatory decisions. I also am concerned that we be able to meet the consumers' need to identify the sources of protein in foods because of sensitivities and allergies. We have become aware that, when protein isolates are formed, the source of the protein becomes more difficult to identify. The consumer has asserted, and the regulatory agencies concur, that the protein source must be identifiable for both health and for other reasons. We feel that industry has a concern in this area as well because reputable processors stand to lose if fraudulent practices are permitted.

REFERENCE

Vanderveen, J. E,. Boehne, J. W., and Adkins J. S. (1977) Cereal Foods World 22, 209-212.

MEASURING PROTEIN QUALITY OF FOODS

F. H. Steinke

When considering the issue of protein quality in foods, we must keep in mind that we are primarily interested in protein quality for human nutrition. Animals other than man or micro-organisms are used to evaluate protein quality because they are models for the human. The results from these test models must be interpreted on the basis of how closely they correlate with the human in protein and amino acid requirement.

Most measurements of "protein quality" measure the content of a single essential amino acid in the food protein relative to the requirement of the species being used as a test model. The present methods of evaluating protein quality of food proteins depend on an animal other than the human to establish the value of a protein for the human population. More specifically, the laboratory rat has been established as the animal model for testing protein quality. The obvious advantages of the rat as a model are:
1)   availability at a low cost;
2)   adaptability to highly refined diet;
3)   well established dietary requirements; and
4)   ability to maximize differences in amino acid content due to a high requirement for essential amino acids.

Animal models, and the rat in particular, have several disadvantages. Animals in most model systems grow at a considerably faster rate than even the most rapidly growing human at any age period. Therefore, the proportion of the amino acids utilized for growth versus maintenance is much greater in the animal models. Laboratory animals have more hair and thus have a somewhat different body composition. The faster growth rate coupled with a different body composition results in an amino acid requirement for the model animal different than the human population to which we are trying to relate the results of the assay methods.

Another inconsistency is the use of casein as the standard reference protein in the evaluation of protein quality. Few

evaluations of casein have been made with human subjects to
determine its protein value relative to other protein
sources. The amino acid composition of casein does not correspond to amino acid pattern requirements of humans; therefore, using casein as a reference is questionable. Casein
does not maximize growth rate for the test animal when compared with egg or good quality lactalbumin, for casein is
limiting in sulfur amino acids. The responses obtained in
protein quality studies with rats or other species using
casein are directly related to the sulfur amino acid content
of the casein. The relative responses are therefore proportional to the animal's requirement for sulfur amino acids. It
is difficult to rationalize the meaning of a protein quality
study when a lysine limiting protein is compared with a reference protein which is limiting in sulfur amino acids.

Evaluating proteins directly with humans to establish a
protein's value as a source of amino acids and protein is one
obvious answer to the questions and problems discussed above.
While theoretically it is possible to conduct these types of
assays, on both an economic basis and due to the limited number of subjects available for testing programs, it is impractical to conduct human feeding studies with a large number of
proteins and products. However, direct human evaluations are
essential to establish amino acid and protein requirements of
humans and to establish the responses of representative protein sources as has been done by Fomon, Torun, Scrimshaw, and
Young with soy protein.

An alternative to direct human evaluations is to use the
human amino acid pattern requirements to develop a scoring
system that would serve as the basis for determining the value
of a food protein for the human population. This would have
the advantage using the target population as the model system
without having to individually feed each protein and combination of food protein. The amino acid score or patterns published by the Food and Nutrition Board (FNB) (1974) and the
FAO/WHO (1973) are shown in Table 1 with a comparable score
derived from the NRC Nutrient Requirement of Laboratory Rats
(1972) for amino acids and protein.

The amino acid pattern requirement for the rat is considerably different than either of the human amino acid pattern
requirements, illustrating the basic inconsistency of using
the rats to measure protein quality for humans. In particular, the sulfur amino acids, lysine, histidine and arginine
requirements differ greatly as a percentage of the protein
requirement. The rat has a specific requirement for arginine
whereas the human apparently does not. Thus, proteins low in
arginine could give a low value for protein quality with the

Table 1. Comparison of Amino Acid Requirement Patterns

Essential Amino Acids	Amino Acids as % of Protein		
		Human	
	Rat	FNB	FAO/WHO
Histidine	2.5	1.7	1.4*
Isoleucine	4.6	4.2	4.0
Leucine	6.2	7.0	7.0
Lysine	7.5	5.1	5.5
Total Sulfur A.A.	5.0	2.6	3.5
Total Aromatics	6.7	7.3	6.0
Threonine	4.2	3.5	4.0
Tryptophan	1.25	1.1	1.0
Valine	5.0	4.8	5.0
Arginine	5.0		

*Infant

rat but have no significance for the human. The greatest differences in amino acid pattern requirements between rats and humans is in the sulfur amino acids, with the rats having a value twice that of the humans based on the FNB pattern. The rat, therefore, selectively discriminates against proteins with lower sulfur amino acid content. The differences in lysine requirements are not as great as the sulfur amino acids requirement, but are still substantially lower, with human requirement being 68% of the rat requirement.

To further illustrate this point, the amino acid scores of some representative proteins based on amino acid analyses conducted in our laboratory are listed in Table 2 comparing the rat pattern with the FNB human pattern. The scores are computed by comparing the amino acid content of the test protein as a percentage of protein (N x 6.25) with the amino acid pattern on Table 1. The score is the proportion of the most limiting amino acid of the test protein to the requirement pattern. The scores using the rat pattern were all at or below 100 indicating that all proteins were suboptimal for the rat with the possible exception of egg protein. The same proteins compared with the human pattern gave a much higher score, with the animal protein and isolated soybean protein having scores of 100, while oat and wheat flour had lower values. Clearly, the rat discriminates against proteins with lower sulfur amino acid and lysine contents.

Rather than needing a more rapid rat assay or a more rapid *in-vitro* assay which correlates with the present PER, slope ratio or Net Protein Ratio assay, a method is needed to evaluate protein quality relative to the human requirement. The methodology for amino acid analysis of the food proteins is

Table 2. Amino Acid Score Based on Rat and Human Requirements

PROTEIN SOURCE	RAT	HUMAN[1]
Casein	71	100
Egg Whole	95	100
Lactalbumin	65	100
Beef	90	100
Pork	87	100
Isolated Soy Protein	58	100
Sesame Flour	36	54
Oat Flour	52	77
Wheat Flour	33	49

[1]*Based on Food and Nutrition Board pattern*

available at a relatively low cost and can be used to establish amino acid content. The comparison of the amino acid content of food proteins with the human amino acid pattern is a valid basis for equating proteins to human needs.

Additional consideration must be given to the digestibility of the protein and the availability of amino acids. The amino acid content of the food protein should be adjusted for the digestibility of protein by some direct measurements. There are obviously a number of species which can be used to establish digestibility, including the rat and pig. Digestibility values obtained by either direct or indirect methods using appropriate markers are much simpler to obtain than protein quality measurement. Digestibility measurements can also be estimated by *in-vitro* methods using enzymatic digestion or microbial utilization. The combination of an amino acid score using both amino acid analysis and digestibility would have the advantage of more rapid analysis and greater direct validity with respect to the human protein needs than the traditional procedures.

An alternate procedure for measuring protein quality which relates more directly to humans can be devised by using the amino acid pattern of the human to modify the interpretation of the results of assays based on rat growth. Usually the highest protein efficiency ratio (PER) obtainable with the rat is 3.2 based on whole egg or good quality lactalbumin. Within the parameters of the PER testing procedures this indicates that all the essential amino acid requirements have been met for the rat. To equate this response of the rat with that of the human, the amino acid requirement of the human can be calculated as a percentage of the rat requirement and this value multiplied by the highest PER for the rat of 3.2 when the ami-

Table 3. *Calculated Minimum PER Values Based on Ratio of*
Human to Rat Amino Acid Pattern

First Limiting Amino Acid	Minimum Satisfactory PER (Adjusted)
Histidine	2.2
Isoleucine	2.9
Leucine	3.2
Lysine	2.2
Total Sulfur A.A.	1.7
Total Aromatic A.A.	3.2
Threonine	2.7
Tryptophan	2.8
Valine	3.1

no acid requirement is fully met. Each protein can be evaluated based on the first limiting essential amino acids determined by amino acid analysis. The minimum satisfactory PER
for a protein limiting in sulfur amino acids would be different from that of proteins limiting in lysine but each can be
based on the relative requirement of the human to the rat requirement for the same amino acid.

The minimum satisfactory PER (MS PER) for sulfur amino
acids would be calculated from the rat requirement and FNB
pattern given in Table 1 as follows:

M.S. PER = maximum adjusted PER for rat x

$$\frac{\text{Sulfur amino acid req. of human as \% of protein}}{\text{Sulfur amino acid req. of rat as \% of protein}}$$

$$= 3.2 \text{ x } \frac{2.6}{5.0}$$

$$= 1.66$$

The minimum satisfactory PER value for the other essential
amino acids are given in Table 3. This procedure would have
the advantage over the present PER methodology since it would
be correlated directly to the human essential amino acid requirement but allow *in-vivo* measurement. While the PER procedure is used here as an example for equating animal growth
assays and human amino acid requirements, the same interpretations can be applied to other animal assay procedures.

However, of the two methods proposed, the amino acid chemical score and digestibility evaluation is preferable due to
the potentially more rapid and economic assay. Either method
would be an improvement over the present method of determining
protein quality of human food products.

REFERENCES

FAO/WHO Ad Hoc Expert Committee (1973). Energy & Protein Requirements, FAO, Rome.

Food and Nutrition Board (1974). Recommended Dietary Allowances, NAS/NRC, Washington, DC.

NAS/NRC (1972). Nutrient Requirement of Laboratory Animals, No. 10, Washington, DC.

Scrimshaw, N. S. and Young, V. R. (1978). Quality of Soy Protein - Adults. This volume.

Torun, B. J. (1978). Nutritional Quality of Soybean Protein Isolates: Studies in Children of Preschool Age. This volume.

A CRITICAL SUMMARY OF A SHORT-TERM NITROGEN
BALANCE INDEX TO MEASURE PROTEIN QUALITY IN
ADULT HUMAN SUBJECTS

R. Bressani, D. A. Navarrete, L. G. Elias, and J. E. Braham

The evaluation of protein quality in human subjects is a
difficult problem mainly due to:
a)   Continuous changes in the metabolic state taking place
while attempting to measure the response of body protein
metabolism to variable dietary intakes;
b)   The determination of energy requirements;
c)   The variability in the efficiency of protein utilization
by individuals in different physiological states, and;
d)   Interpretation of results in relation to the meaning of
theoretical models and evaluation for practical purposes.
Because of these and other possible problems the cost of
such assays is quite high, representing a significant con-
straint to the establishment of standard experimental conditions to obtain reproducible and constant responses to specific protein sources.
Presently, various approaches are utilized to measure protein quality. However, one which overcomes most of the difficulties is a multiple point assay originally known as the
Nitrogen Balance Index (Allison, 1945; Bressani and Viteri,
1971; Bressani et al, 1973). This assay relates through a
regression equation, the relationship between nitrogen intake
or nitrogen absorbed to nitrogen retained, which according to
theory should be linear in the region below and slightly above
nitrogen equilibrium. The theoretical line is shown in Figure
1, which also shows the regression equation NB + K(NA) - Neo,
where NB signifies nitrogen balance and NA nitrogen absorbed.
The value K is equivalent to the efficiency of utilization of
the protein under study and is the slope of the line; and Neo
represents the total excretion of nitrogen of metabolic and
endogenous origin from feces and urine.
We believe that in theory, the point enclosed by circle 1,
in Figure 1, represents the sum of metabolic fecal and of

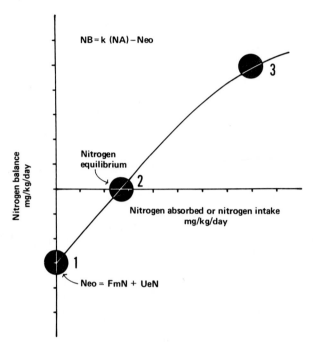

FIGURE 1. The basic principle of the nitrogen balance index.

endogenous urinary nitrogen, obtained by feeding a nitrogen-free diet. Only rarely does the extrapolated value (i.e., the intercept) from the regression equation coincide with that derived from the actual feeding of a nitrogen-free diet. Typical of this are results reported for egg and other proteins in recent publications (Young et al, 1977; Bressani, 1977). The reasons for such discrepancies are not known, but could be due to metabolic causes, to methodological technique, or both. In any case, it is an area which must be investigated. In Figure 1, circle 2 is the level of nitrogen absorbed to meet maintenance requirements. This level is determined mainly by the amino acid limiting the efficiency of utilization of the protein or, in other words, the most limiting amino acids needed to meet the maintenance requirement. These amino acids may be different in adults from those in young growing subjects. In theory, this point must be different in terms of nitrogen, for proteins of different quality.

If food proteins are limiting in different essential amino acids, but proportionally to the same extent, the amount of nitrogen retained at a given level of nitrogen intake should be the same for all of them. Furthermore, in theory, the minimum amount of nitrogen retained under normal and acceptable experimental conditions for a high quality protein, would take place when the N balance line crosses the N absorbed line at a 45° angle. For high quality proteins, that is, for those with a high efficiency of utilization, the amount of nitrogen required for nitrogen balance should be numerically only slightly higher than the value for total metabolic, fecal and urinary endogenous nitrogen excretion. If the above is true, this point is quite meaningful, in terms of expressing protein quality, and can be used as a reference point to calculate relative protein quality values.

Circle 3 represents excess nitrogen intake, which is wasted, or utilized with a low efficiency of nitrogen utilization. This point is believed to be determined by overall absolute amino acid level in which the limiting amino acid of the protein no longer plays an important role in establishing true biological value.

The nitrogen balance values from area 3 are not sensitive to protein quality and in the past have led to erroneous conclusions with respect to the significance of biological value of proteins in human adults. The biological effects, if any, from such N retentions from poor quality proteins are not known.

In our opinion, the meaning of these points and the correlation between theoretical and experimental results should be assessed; a limitation, however, is the time required to run a test. Customarily, in nitrogen balance studies with humans, the length of each feeding period is of 10 to 14 days duration. The number of individuals needed for each test is a complicating factor. To be able to solve some of these problems, as well as to be able to develop a protein assay technique, we studied the possibility of reducing the experimental period, after obtaining promising results with adult dogs, as experimental animals (Bressani et al., 1978; Navarrete et al, 1977; Bressani et al, (in press); Navarrete et al, (manuscript in preparation). The experimental conditions established for adult human subjects are shown in Table 1.

Subjects between 20 and 35 years of age are chosen on the basis of good general health as judged by medical and biomedical parameters, ongoing dietary intake and physiological functions. In various experiments between 5 and 12 subjects were fed each diet protein. Once in the experimental run they are fed from 40 to 45 kcal/kg body weight/day, of which 30% of the calories are derived from fat. A constant and adequate water

Table 1.   Basic Procedure Followed For Short Time Nitrogen
           Balance Index

Subject age:	20 - 35 years old
Calories/day:	40 - 45 kcal/kg
Water intake:	Adequate and Constant
Low nitrogen diet:	10 - 15 mg/kg/day

Protein intake g/kg/day	Days on balance
0.0	3
0.2	2
0.4	2
0.6	2

intake is provided. The order of levels of protein feeding
are shown in the lower part of the Table. A multivitamin cap-
sule is given daily to all subjects. With the aid of markers,
feces are quantitatively collected, as well as urine, for a
total period of two days.

Results using this modified technique as well as the conventional method are summarized in Table 2. The relationship
between nitrogen intake (NI) to nitrogen balance (NB) is expressed as the regression coefficient, b. Although the number
of conventional runs is small in comparison with the short-
term approach, agreement between the two is quite good. Furthermore, the values found by the short-term assay agree with
those reported by other workers (Bressani, 1977; Navarrete et
al, 1977; Bressani et al, (in press); Navarrete et al, (manuscript in preparation). Of the group, beans show the poorest
quality, needing around 116 mg N/kg/day to attain equilibrium.
Without considering egg, the best quality protein is milk,
given an intake of around 80 mg/kg/day for nitrogen equilibrium. Table 3 summarizes the same information; however, the
relationship here is between NA and NR.

Based on the coefficient of regression which varies from
0.70 for casein to 1.06 for milk, the ten proteins so far
tested are of high quality. Egg protein has been very erratic
for reasons not yet known; however, the values found, except
for the 1.02, fall within the range reported by other workers.
The high values are probably due to prolonged time on the
nitrogen-free diet. It is our intention to test intermediate
and low quality proteins once the experimental conditions are
well established. The values for the short assay are similar
to those obtained from the conventional method in our laboratory.

Table 2. Relationship Between Nitrogen Intake (NI) and
Nitrogen Balance (NB) as Obtained From a Short
Term Nitrogen Balance Index In Adult Humans

Protein Source	Short Assay NR = a + b (NI)		r	Conventional Assay NB = a + b (NI)		r
	a	b		a	b	
Egg	−55.21	+ 0.86	0.88	−57.58	+ 0.70	0.89
	−63.64	+ 0.57	0.78			
Milk	−78.80	+ 0.91	0.84	−72.34	+ 0.82	0.92
	−73.61	+ 0.98	0.92	−70.38	+ 0.77	0.91
	−81.30	+ 1.03	0.84			
	−70.60	+ 1.00	0.82			
Casein	−60.28	+ 0.64	0.90			
Soybean-TVP	−65.73	+ 0.68	0.92			
Soybean Isolate	−72.19	+ 0.83	0.90			
Beef	−74.40	+ 0.87	0.95			
Soybean-Beef	−79.90	+ 0.87	0.89			
Figurines	−75.44	+ 0.90	0.91			
Corn/Beans	−86.81	+ 0.89	0.89			
Beans	−62.69	+ 0.54	0.75			

Table 3. *Nitrogen Balance Index Values Obtained in Adult*
Subjects at Incap

Protein	Short Assay			Conventional Assay		
Source	NB = a + b (NA)		r	NB = a + b (NA)		r
	a	b		a	b	
Egg	−59.45	+ 0.67	0.78	−39.65	+ 0.57	0.81
	−51.58	+ 1.02	0.95			
	−54.91	+ 0.37	0.52			
Milk	−54.49	+ 0.91	0.93	−57.21	+ 0.93	0.97
	−47.08	+ 0.97	0.91	−56.16	+ 0.88	0.94
	−56.53	+ 1.06	0.73			
	−41.89	+ 0.88	0.84			
Casein	−48.42	+ 0.70	0.92			
Soybean−TVP	−54.21	+ 0.77	0.95			
Soybean Isolate	−56.29	+ 0.91	0.93			
Beef	−54.88	+ 0.86	0.97			
Soybean−Beef	−61.76	+ 0.91	0.94			
Figurines	−52.05	+ 0.83	0.91			
Corn−Beans	−63.81	+ 0.95	0.92			
Beans	−57.44	+ 0.81	0.82			

Various aspects of the results presented should be indicated, as shown in Table 4. The numerical experimental figures for total endogenous N excretion from the short method
regression equations range from 42 to 64, with an average of
54. The average value as found in the literature is 50, of

Table 4. *Summary of Values Derived From Regression Equations*
From the Short-Term Nitrogen Balance Index Assay

	Equations:	NB = Neo + K (NI)	(1)	
		NB = Neo + K (NA)	(2)	

Parameters		*Short-Term*	*Conventional*	*Literature*
Neo (1)	Range	55.2 - 86.8	57.6 - 72.3	-
	Average	71	65	-
Neo (2)	Range	41.9 - 63.8	56.2 - 57.2	-
	Average	54	57	50

which 39 came from urine and 12 from feces. We believe that a good NBI assay should show a Neo value close to the theoretical figure.

The results obtained by the short assay also confirm results reported by other workers using the conventional approach as shown in Table 5. In this case the values of nitrogen intake and of nitrogen absorbed for nitrogen equilibrium are compared. The agreement is high between the results from the conventional and short-term assays as presently carried out within INCAP and also with values from other laboratories. Further evidence supporting the potential of the short-term method is presented in Table 6, showing protein intake values for nitrogen equilibrium for 4 protein sources assayed by the conventional and short method. Although the agreement is quite high, two points are of concern:
a)   the short technique has a tendency to give higher values for efficiency of N utilization, and
b)   the point for nitrogen equilibrium is lower.

The first situation shown in Figure 2 is due to the fact that when protein intake is increased and maintained at the same level, the NR during the adaptation period is more negative at below maintenance requirements and more positive above maintenance requirements than the NR for the balance period. Therefore, the slope using the initial retention values at a fixed N intake are somewhat higher than if the following NR values are used to calculate the regression equation. Therefore, the nitrogen equilibrium point also changes, becoming smaller. If efficiency of utilization (the slope) on nitrogen intake required to meet nitrogen balance of a test protein is expressed as percentage of a standard protein, there is probably no problem in interpretation, since the same situation will prevail for the reference protein at all nitrogen intakes.

Table 5.   *Some Comparative Values of Nitrogen Index (NI) and Nitrogen Absorption (NA) for Nitrogen Equilibrium (NE) for Various Proteins*

Protein	Method	Author	NI for NE	MA for NE
			mg/kg/day	
Egg [1]	Conventional	Calloway & Margen	–	68
Egg	Conventional	Young et al '73	88	73
Egg	Conventional	Inoue et al '77	–	100
Egg	Conventional	INCAP	82	69-83
Egg	Short	INCAP	64-112	50-89
Milk	Conventional	Bricker et al '45	–	60
Milk	Conventional	INCAP	88	61-64
Milk	Short	INCAP	75-86	48-60
Casein	Short	INCAP	94	69
Soy TVP	Short	INCAP	97	70
Soy Isolate	Conventional	Young et al '73	107	–
Soy Isolate	Short	INCAP	87	62
Meat	Short	INCAP	85	64
Soy/Meat	Short	INCAP	92	68
Figurines	Short	INCAP	84	63
Corn/Beans	Short	INCAP	97	67
Beans	Short	INCAP	116	71

[1]See Footnotes Table 2.

Table 6. Protein Intake for Nitrogen Equilibrium, g/kg/day

Protein Source	Conventional [1,2]	Short-Term
Soy Isolate	0.67	0.54
Milk	0.63	0.62
50/50 Beef/Soy	0.59	0.57
Beef	0.64	0.53

[1]*Scrimshaw and Young. This conference.*
[2]*Values corrected assuming 5 mg integumental and miscellaneous losses.*

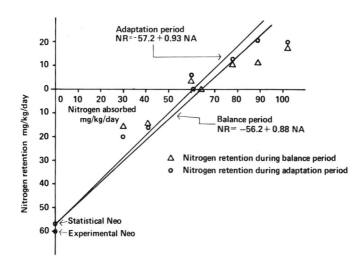

FIGURE 2. Nitrogen balance index of milk with adult human subjects. (Conventional Assay).

Up to the present time, only high quality proteins have been tested and it appears, from the data obtained, that there is rather good agreement between theory and actual results. It is of practical significance to discover if the same applies to protein of low quality with single or double amino acid deficiency. These are other aspects which are under study in our laboratories, as well as other aspects. These are:

a) The effect of protein level and length of intake during adaptation period before the test;

b) The length of the feeding period of a nitrogen-free diet (NFD) before protein feeding (no longer than four days);

c) The number of protein levels to be fed below and above the equilibrium point.

The results in experimental animals, although preliminary, so far indicate that previous dietary protein level before the test does affect the slope and intercept. The length of NFD feeding, which has a more significant influence, also influences the slope and intercept.

Further research is needed to establish the correct experimental conditions to assay protein quality in humans, and the short-term approach may be useful for such a purpose, as well as for establishing protein quality values and requirements.

REFERENCES

Allison, J. B and Anderson, J. A. (1945). *J. Nutr.* *29*: 413.

Bressani, R. (1977). "Evaluation of Proteins for Humans." (C. E. Bodwell, ed.). Chapter V, pp. 81-118. The AVI Publishing Company, Inc., Westport, CT.

Bressani, R.; Elias, L. G.; Olivares, J.; and Navarrete, D. A. (1978). *Archivos Latinoamer. Nutr.* In press.

Bressani, R.; Navarrete, D. A.; Loueiro de Daqui, V. A.; Elias, L. G.; Olivares, J.; and Lachance, P. A. (1978). *J. Food Sci.* Submitted for publication.

Bressani, R. and Viteri, F. (1971). "Metabolic Studies in Human Subjects." Proceedings of the Third International Congress. SOS/70 pp. 344-357. Washington, DC.

Bressani, R.; Viteri, F.; and Elias, L. G. (1973). "Protein in Human Nutrition" (J. W. G. Porter and B. A. Rolls, eds.). Chapter 19, pp. 293-316. Academic Press. London and New York.

Navarrete, D. A.; Elias, L. G.; and Bressani, R. (1978). Manuscript in preparation.

Navarrete, D. A.; Loueiro de Daqui, V. A.; Elias, L. G.; Lachance, P. A.; and Bressani, R. (1977). *Nutr. Reports Intern. 16:* 695–704.

Young, V. R.; Rand, W. M.; and Scrimshaw, N. S. (1977). Cereal Chemistry 54:929–948.

COMPARISON OF HUMAN
AND ANIMAL STUDIES AT NEBRASKA[1]

Constance Kies

During this meeting, reviews of numerous methods for measuring protein quality of soy products have been presented.
The concept of the existence of a perfect method for protein
quality evaluation or the possibility of its development is
undoubtedly false. In evaluation of methods, each individual's opinions are colored by his own circumstances and expe-
riences. All of the methods offer favorable advantage/disad-
vantage ratios for protein evaluation under particular conditions of place and time, but similarly under a different
environment, they might be classified as poor. To aim for
perfection in methodology development is admirable but to
expect to achieve it is naive.

The Food and Nutrition Department at the University of
Nebraska is not and has not been in the business of development of protein quality evaluation methodology; rather application and, to a lesser extent, evaluation of methods already
known has been an emphasis. Because of peculiarities and
uniqueness of funding, available facilities and experimentor
expertise, bioassays of proteins using humans has been a specialty area. Because of this, I have been most interested in
those assays presented here involving that "experimentation
model."

To suggest that all protein-rich products should be monitored for protein quality using human bioassay techniques is
impractical and possibly immoral. However, use of human bioassays to standardize other techniques for product testings is
desirable. The functions of digestion, absorption, and metab

[1]Paper No. 5562, Journal Series, Nebraska Agricultural
Experiment Station. Funds for this project were provided by
Nebraska Agricultural Experiment Station Project No. 91-007.

olism of protein and amino acids vary among species and, to a
lesser extent, within species and even within the same individual of the species under different physiological, psychological and environmental conditions. Establishment of
methods which will accurately predict performance is difficult
in and of itself and is doubly difficult when answers are
sought for human performance.

Whether or not one is discussing human bioassays, animal
bioassays, microbioassays or even chemical bioassays, the
question of single point assays versus multiple point assays
is an important one. From the purely scientific standpoint,
the virtues inherent in multiple point assays simply cannot be
argued. However, there are pertinent arguments from the
application/practical approach. In animal and human bioassays
costs are greatly magnified in multiple point assays because
of increased requirements of product to be tested, time,
equipment, labor, human or animal models. In the case of
human subjects, the lengthening of experimental time required
for multiple point assays may place the subject at greater
risk; thus, raising the question of moral justification if
data obtained is not of substantially greater value.

In the UN-L (University of Nebraska-Lincoln) laboratory,
we have done a few single and multiple point assays on soy
products using several different approaches (Kies and Fox,
1971; Kies and Fox, 1973; Korslund, 1973; Vemury, et al, 1976;
Vemury, et al, 1978; Ranum, et al, 1977).

In this series, one project involved the comparison feeding of TVP (Textured Vegetable Protein, an extruded defatted
soy meal product produced by the Archer Daniel Midlands
Company), a methionine enriched TVP to 9 adult men at 2 levels
of nitrogen intake (4 and 8 g N/subject/day) during separate,
randomly arranged periods of 7 days each (Kies and Fox,
1971). Mean nitrogen balances of subjects while fed 4.0 g N
from TVP, methionine-enriched TVP, and beef were -0.70, -0.45
and -0.30 g N/subject/day respectively while those of subjects
on the same diets at the 8.0 g N intake level were +0.78,
+0.72 and +0.74 g N/subject/day, respectively.

These figures can be interpreted in several ways. Looking
at the 4.0 g N intake figures, one could say that soy protein
is definitely inferior to beef protein but can be improved by
methionine supplementation. If one looks at the 8.0 g N
intake, one could say that at "normal" protein intake levels
soy, methionine enriched soy, and beef all meet human protein
needs and no difference in protein quality was demonstrated.
(One, of course, cannot say that the 3 products were proven to
be equal.) Protein quality differences can be measured by
bioassay only at inadequate intake levels. In working with
food products supplying all essential amino acids even if not

in idealized ratios, quantity can, to a large extent, compensate for quality. If one determines slope ratios using these
points, soy would actually have the steepest slope; thus, the
apparent highest rather than poorest value. This would constitute misuse of the method, of course, since undoubtedly the
beef diet would allow achievement of maximum positive nitrogen
balance at a lower level than will the soy - something not
illustrated by this study.

Another problem which is illustrated by this study that
has not been fully discussed in this symposium is that of
interchangeable use of the terms nitrogen and protein. To do
comparisons at equal levels of nitrogen although customary in
both human and small animal bioassay procedures is not the
same as comparisons at equal levels of protein. Not all proteins are 16% nitrogen; hence, the conversion factor of 6.25
for calculating protein from nitrogen content is not always
accurate. This difference puts soy protein at a particular
disadvantage in comparison to beef, milk, or casein standards
since a more accurate conversion factor for soy is about 5.71
based on its nitrogen content relative to protein while that
for beef is 6.25. Thus, in the aforementioned study at the
4 g nitrogen intake, actually only 22.8 g of soy protein was
fed in comparison to 25.0 g of beef protein. This then places
the soy in a falsely disadvantageous position in determining
protein quality. By the same reasoning, soy is placed in a
falsely advantageous position in calculating protein quantity
using the 6.25 factor.

With new developments of instrumentation and methods,
amino acid profiles for food products can be obtained with
greater ease than formerly. Merlin Vemury, as part of her
doctoral research in our laboratory, worked with a series of
beef/soy snack type products which were being considered for
production by the Fairmont Foods Company. (Vemury, 1976,
Vemury, et al, 1978). Amino acid profiles were determined on
these products and new nitrogen to protein conversion factors
were calculated based on these profiles. Products and the
determined conversion factors were as follows: regular beef
jerky = 6.25; 10% TVP + 10% beef jerky = 6.15; 10% soy concentrate + 10% beef jerky = 6.14; 14% soy concentrate + 6% beef
jerky = 6.13; regular beef sausage = 6.25; 20% TVP sausage =
6.06; 30% TVP sausage = 5.99. In formulating rations for
doing PER measurements in rats these conversion factors were
employed so that rations contained equal levels of protein at
the 10% level rather than equal levels of nitrogen. By this
system, the soy/beef blended jerky products received relatively high marks - values being as follows: regular beef
jerky = 2.40; 10% TVP + 10% defatted beef jerky = 2.95; 10%
soy concentrate + 10% defatted beef jerky = 3.01; and 14% soy

concentrate + 6% defatted fat beef jerky = 2.48. For the sausage products, PER values were as follows: regular beef sausage = 2.48; 20% TVP sausage = 1.77; 30% TVP sausage = 1.76.

Research sometimes overcomplicates relatively simple questions. Often the question asked is simply whether the protein quality of one product is better or worse than another or will a particular treatment improve or damage protein quality. Thus, we have run a series of studies on whether or not chemical, small animal, and human bioassays will give essentially the same simple relative answers. Regrettably for me in this symposium, these tests have been run on different wheat varieties rather than on soy products but the general approach holds true regardless of product under investigation.

On the first study 3 wheat grain materials were Gage, Scout 66 and Scoutland grains having a protein content of 16.8, 15.8, and 14.7%, respectively (Kies, et al, 1972). Lysine contents of the grains on a percent protein basis were 4.85, 3.06, and 3.11, respectively, and on a percent grain basis were 0.48, 0.48, and 0.46, respectively.

Throughout this study, ranks were assigned from a 1 (best) to 3 (poorest) basis for whatever characteristic or type of data was being evaluated. Using this approach, on the characteristic of protein content, Gage was ranked 1, Scout 66 ranked 2, and Scoutland ranked 3; on the basis of lysine on a percent protein basis, Scoutland was ranked 1, Scout 66 was ranked 2, and Gage was ranked 3; on the basis of lysine content on a percent grain basis, Scout 66 was ranked 2.5, Gage was ranked 2.5, and Scoutland was ranked 3.0.

Using this approach, PER values were determined in mouse bioassays and nitrogen balances of humans were determined using diets containing equal levels of nitrogen and equal levels of protein.

In reference to protein quality the following rank values were obtained in the test used: Lysine content: 3 = Gage, 2 = Scout 66, and 1 = Scoutland. Mouse PER values by both 7% and 10% protein rations: 3 = Gage, 2 = Scout 66, and 1 = Scoutland. Human bioassay (measurement at 4.0 g N intake): adult subjects -- 2.5 = Gage, 2.5 = Scout 66, and 1 = Scoutland; child subjects -- 3 = Gage, 2 = Scout 66, and 1 = Scoutland.

Thus, by chemical assay, by small animal bioassay, and by human bioassay, simplified answers to protein quality were nearly identical. In this project, other comparative assays relative to protein quality/quantity were also done. These rank values were different from the rank values for protein quality but also good correlation was found among chemical assays, small animal bioassays, adult human bioassays, and adolescent children bioassays.

Genetically and chemically, wheat grain from the wheat varieties Gage, Scout 66 and Scoutland are closely related. In a second, more elaborate study, similar comparisons as above were made between whole ground flours and extracted flours from 5 less closely related wheat lines: Atlas 66, Centurk, Naphal, Bezastaya, and commercial wheat (Kies, 1975). Since these materials were less alike chemically and genetically than those used in the first study of this project, it was assumed that demonstrated differences would be even more clearly cut. In fact, the reverse was true. Protein quality in human bioassays on a rank basis could not be completely predicted either from chemical assays or from rat or mouse bioassays.

Digestibility as related to fiber content and to location of lysine within the wheat kernel seemed to be of primary importance. In the first study, these factors were of minimal importance because the test grains were closely similar to these factors. However, the reverse was true in the second and these factors overrode some of the assumed primary considerations of protein quality such as amino acid proportionality patterns.

Digestibility and availability related to protein quality has been discussed and methods for its measurement presented. While the aforementioned wheat study emphasizes the need to take digestibility into consideration, it is also wise to remember that the highest degree of digestibility and thus availability is not always most advantageous for metabolic usage. It has been known for years that purified amino acid diets which are really predigested and 100% available are not utilized as well as intact proteins having the same amino acid proportionality patterns (Kies, 1960; Nasset, 1957; Rose, et al, 1954; Shortridge, et al, 1961). One explanation is that the amino acids are too quickly absorbed and body mechanisms for utilization are simply overwhelmed. Thus, there are limits to the benefits to be obtained from increasing digestibility of protein.

In conclusion, I must inject a negative thought. For several years I worked with protein quality and found it a highly interesting topic from a research standpoint. In the development of soy products, I wonder if questions of protein quality have been overworked. Soy protein quality really is quite good. Rats, mice, pigs, cattle, chickens, and humans have no problems existing, growing, and reproducing on protein supplied by soy. I was happy to see other nutrients touched on in this symposium. Whether one is discussing soy products or other products, nutritional value is not merely a question of protein quality. A balanced approach to nutritional evaluation is an idea to be encouraged.

REFERENCES

Kies, C. (1960). Studies in urinary nitrogen excretion. M.S. Thesis, University of Wisconsin.

Kies, C. (1974). *Cereal Sci. Today 19:* 450-469.

Kies, C. (1975). Comparative protein nutritive values of whole ground and white flours from five wheat varieties for humans. Paper. Milling and Baking Division, Am. Assoc. of Cereal Chemists, Spring Technical Conference.

Kies, C. and Fox, H. M. (1971). *J. Fd. Sci. 36:* 841-845.

Kies, C. and Fox, H. M. (1972). Comparative protein quality as measured by human and small animal bioassays of three lines of winter wheat. Paper presented at the Am. Assoc. of Cereal Chemists, Annual Meeting.

Kies, C. and Fox, H. M. (1973) *J. Fd. Sci. 38:* 1211-1213.

Korslund, M., Kies, C., and Fox, H. M. (1973). *J. Fd. Sci. 38:* 637,638.

Nasset, E. S. (1957). *J. Nutrition 61:* 171.

Ranum, P., Kies, C. and Fox, H. M. (1977). *Nutri. Reports Intern. 16:* 425-437.

Rose, W. C., Coons, M. J. and Lambert, C. F. (1954). *J. Biol. Chem. 210:* 331.

Shortridge, L., Chao, K. C., Kies, C. and Reyholds, M. S. (1961). *Federation Proc. 20:* 8.

Vemury, M. (1976). Protein nutritional value of methionine limited foods as influenced by product formulation and supplementation. Ph.D. Thesis. University of Nebraska - Lincoln.

Vemury, M., Kies, C. and Fox, H. M. (1976). *Federation Proc. 35:* 743.

Vemury, M., Kies, C. and Fox, H. M. (1978). *Reports Intern. 17:* 417-430.

HUMAN VERSUS ANIMAL ASSAYS[1]

C. E. Bodwell

Traditionally, values from animal assays have been used as general estimates of protein nutritional value for humans. In the early studies conducted to evaluate the relation between estimates of protein nutritive value for humans and for rats, Biological Value as determined in rats agreed well with Biological Value as determined in humans. An unrecognized factor at the time the early studies were conducted was the effect of protein intake levels on estimates of Biological Value. In the human studies, protein sources expected to be of high quality were fed at low levels while sources expected to be of low protein nutritive value were fed at high levels. We now know that this approach was invalid. Data from more recent studies are more useful for evaluating the relationship between protein nutritive value as estimated by rat assays and by assays with human subjects. Studies in which the same protein sources were evaluated in both rats (or other animals) and in humans have been recently reviewed (Bodwell, 1977) in greater detail than is possible in this brief paper. However, it is possible to present data from those rat and human studies which have contributed the bulk of the comparative data available.

De Meyer and Vanderborght (1961) evaluated several protein sources in children. Of these, four of the same preparations were evaluated in rat studies by Bender (1961, 1973).

[1]Opinions expressed are those of the author and do not reflect the official views of any department or agency of the Federal Government; mention of a Trademark or other proprietary product does not constitute a guarantee or warranty of the product by the U.S. Department of Agriculture and does not imply its approval to the exclusion of other products that may also be suitable.

Summarized data are shown in Table 1. Values obtained by human and rat assays were in agreement for the sesame flour. However, for the other protein sources the relationship was less satisfactory.

Relative protein nutritive values as estimated by the Nitrogen Growth Index (the slope of the regression line relating growth to the intake of nitrogen at several intake levels of nitrogen) in rats and by the Nitrogen Balance Index (the slope of the regression line relating nitrogen balance to nitrogen absorbed) in children are shown in Table 2. Milk, a mixture in which 30% of the protein was derived from cassava and 70% from soy flour, and four INCAP vegetable protein mixtures (compositions given by Bressani, 1975, 1977) were evaluated both in rats and in children. The rat assay estimates were in good agreement with the values from the studies with children for the INCAP Mixture 15, and the IRL mixture. Agreement between the two estimates of the protein nutritive value of the INCAP Mixture 14 was poor and results for the other two sources were somewhat in-between.

Relative Net Protein Utilization values (Egg = 100) as estimated from true digestibility X Biological Value in adult male humans (Bodwell, 1977) and in rats (Hackler, 1977) are listed in Table 3.

Table 1. *Biological Value (BV) or Net Protein Utilization (NPU) of Four Protein Preparations Estimated With Children (3-7 years) and Rats*

Protein source	Protein intake (g/kg body wt/day)	Children[a]		Rats[b]
		BV	NPU	BV
Reference Cow's Milk	0.52-1.49	86-91	79-81	80
Sesame Flour	1.71-3.23	59-62	53-54	58
Peanut Flour	1.55-4.37	54-61	52-57	48
Cottonseed Flour	1.73-5.15	42-62	39-51	65

[a]*From De Meyer and Vanderborght (1961).*
[b]*From Bender (1961, 1973).*

Table 2. *Relative Protein Value (Milk = 100) as Estimated by*
Nitrogen Growth Index in Weanling Rats and by Nitro-
gen Balance Index (Nitrogen Absorbed versus Nitrogen
Balance) in Children[a]

| | Relative protein value | |
Protein source	Rat assay	Children
Milk	(100)	(100)
INCAP Mixture 9	56	67
INCAP Mixture 14	65	85
INCAP Mixture 15	61	64
Cassava - Soy Flour Mixture	86	92
IRL	77	80

[a]*Calculated from Bressani (1975, 1977).*

Table 3. *Relative Net Protein Utilization (Egg = 100) of the*
Same Protein Sources Estimated with Humans (0.4 g
Protein/kg Body Wt./Day)[a] *and Rats*[b]

| | Net Protein Utilization | | |
Protein source	Humans	Rats	(A) - (B)
Spray Dried Whole Egg	(100)	(100)	--
Tuna	93	89	+ 4
Cottage Cheese C	98	91	+ 7
Soy Isolate B	89	66	+23
Wheat Gluten	60	66	- 6
Peanut Flour	93	62	+36

[a]*Bodwell (1977) and unpublished data.*
[b]*Hackler (1977) and unpublished data of L. R. Hackler.*

Agreement between the estimates was reasonably good for
the tuna, cottage cheese and wheat gluten protein sources but
poor for the soy isolate B (*Promine F*) and peanut flour protein sources. Differences in nutritive value as estimated by
relative Net Protein Utilization values in the humans and nutritive value as estimated by Relative Protein Value (RPV) or
Relative Nitrogen Utilization (RNU) rat assays are listed in
Table 4. Both rat assays gave relative values (Egg = 100)
markedly lower than the comparable relative Net Protein Utilization values obtained with the human subjects.

The relationship between Protein Efficiency Ratio (PER)
values for whole egg, milk and seven INCAP protein mixtures
and relative protein value as determined from the regression
coefficients of nitrogen balance vs. nitrogen intake in children (Viteri and Bressani, 1972) is shown in Figure 1 (closed
circles). Little agreement is apparent between the PER values
and protein nutritive value as estimated in the children. The
relative Net Protein Utilization values obtained in four
adults for tuna, cottage cheese, soy isolate B, wheat gluten
and peanut flour are also plotted versus PER values in Figure
1. Agreement is better; however, the range of PER values
(0.50 to 3.0) is much broader than the range in nutritive
value as estimated in the humans.

Table 4. *Differences in Estimated Relative Protein Value
(Egg = 100) of the Same Protein Sources Estimated
in Humans (NPU)[a] and in Rats (RPV, RNU)[b]*

Protein source	NPU estimate (humans) minus RPV estimate (rats)	NPU estimate (humans) minus RNU estimate (rats)
Spray Dried Whole Egg	--	--
Tuna	+19	+23
Cottage Cheese C	+37	+17
Soy Isolate B	+39	+34
Wheat Gluten	+41	+28
Peanut Flour	+56	--

[a]*Bodwell (1977) and unpublished data.*
[b]*Hackler (1977) and unpublished data of L. R. Hackler.*

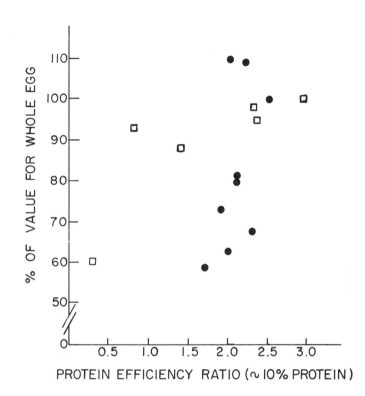

FIGURE 1. Relation between Protein Efficiency Ratio (PER) values and estimates of nutritive value in humans: ● = values for children and corresponding PER values from Viteri and Bressani (1972); □ = values for adults (Net Protein Utilization values expressed as a percentage of value obtained for egg protein) from Bodwell (1977) and unpublished data and corresponding PER values from Hackler (1977) and unpublished data of L. R. Hackler.

The protein nutritive values of two soy isolates (*Supro 620* and *Supro 710*), and milk or whole egg as estimated in children and in young men (in studies described by Dr. Scrimshaw and Dr. Torun previously in this conference) and as estimated by PER assays (unpublished data of F. H. Steinke and D. T. Hopkins) are listed in Table 5.

Based on the regression coefficient in the young men between nitrogen balance and nitrogen intake, the isolate studied has a nutritive value equivalent to about 86% of egg protein. For the children, the estimated nutritive value of

Table 5.   Protein Nutritive Value of a Soy Isolate and
           Reference Proteins as Estimated in Children and
           Young Men by Nitrogen Balance Studies and by
           Protein Efficiency Ratio (PER)[a]

Protein source	Regression coefficient between nitrogen balance and nitrogen intake	Relative value (Egg = 100)	PER (Rats)
Young Men			
Egg	0.50	(100)	3.24
Soy protein isolate[b]	0.43	86	1.65
Children			
Egg	0.58	(100)	--
Milk	0.64	110	--
Soy protein isolate[b]	0.65	112	1.63
Soy protein isolate[c]	0.59	101	1.78

[a]Values for young men from Young et al (1977), for
children from de Godinez (1977), Torun (1978), Viteri and
Alvarado (1971), and Viteri and Bressani (1972), and for rat
assays from F. H. Steinke and D. T. Hopkins (unpublished data).
    [b]Supro 620.   Ralston Purina Company, St. Louis,
Missouri.
    [c]Supro 710.   Ralston Purina Company, St. Louis,
Missouri.

both isolates was higher than estimates for egg or milk protein obtained with children in similar studies. In contrast,
the PER values (corrected values; casein = 2.5) would indicate
a much lower nutritive value for the isolate soy protein than
for the reference egg protein.
    Lastly, relative protein value (casein bread = 100) for 7
protein sources, as estimated in studies at Beltsville by
three different rat assays, are listed in Table 6. Other than
the fact that the values for casein bread were equivalent to
the values for lactalbumin bread, these data indicate differ-

Table 6. *Relative Protein Nutritive Value (Casein Bread =*
100) of Seven Proteins as Estimated by Rat Assays[a]

| Protein source | Assay | | | |
(breads)	RPV	NPR	RNU	Average
Casein	(100)	(100)	(100)	(100)
Lactalbumin	97	100	100	99
Egg white	128	109	109	115
Textured soy protein[b]	75	85	85	82
Soy isolate B[c]	60	76	76	71
Peanut flour	67	67	67	67
Wheat gluten	32	35	34	34

[a]*Calculated from unpublished data (Womack, 1978).*
[b]*Supro 50-4.* *Ralston Purina Company, St. Louis,*
Missouri.
[c]*Promine F.* *Central Soya, Fort Wayne, Indiana.*

ences between protein sources which could be anticipated from
a general knowledge of the protein sources *per se.* The PER
values were likewise generally similar to values which could
be anticipated. However, we have also evaluated the same protein breads at varied protein intake levels in young adult
men. All of the necessary chemical analyses have not been
completed. On the basis of a preliminary survey of the data
using calculated nitrogen intakes (as opposed to intakes based
on analyses of diet composites), however, it would appear that
the rat assay values will generally underestimate the nutritive value of several of the proteins relative to the estimates obtained with the human subjects.

In conclusion, limited data are available from studies in
which the same protein sources have been evaluated in both
rats and humans. However, the data available suggest that
none of the rat assays can be used to consistently and accurately predict protein nutritive value as determined in humans, either children or adults. A need exists for further
comparative studies in which the same protein preparations are
evaluated by rat assays and by assays with humans (preferably
children). Seriously needed data in relation to nutritional
labeling would be obtained if the same protein preparations
used in multiple-intake level studies with humans were routinely assayed by the various rat assays including RPV, PER,

NPR and RNU assays. The lack of a relationship between estimates of protein nutritive value obtained by the PER assay and
values obtained in human assays for specific proteins may also
exist with other assays based on rat growth (including the RPV
assay).[2]

REFERENCES

    Bender, A. E. (1961). *In* "Progress in Meeting Protein
Needs of Infants and Children." Pub. 843, NAS-NRC, pp. 407-
424.
    Bender, A. E. (1973). "Nutrition and Dietetic Foods," p.
165, Chemical Pub. Co., New York, NY.
    Bodwell, C. E. (1977). Cereal Chem. *54*, 958-983.
    Bressani, R. (1975). *In* "Nutrition," Vol. 4 (*Proc. 9th
Intern Cong. Nutr.*, Mexico, 1972). S. Karger, Basel,
Switzerland.
    Bressani, R. (1977). *In* "Evaluation of Proteins for Hu-
mans," (C. E. Bodwell, ed.), pp. 81-118. AVI Pub. Co.,
Westport, CT.
    de Godinez, C. M. (1977). Thesis: "Biological Evaluation
of a Soy Protein Isolate in Preschool Children." University
of San Carlos, Guatemala.
    De Meyer, E. M. and Vanderborght, H. L. (1961). *In*
"Progress in Meeting Protein Needs of Infants and Preschool
Children." Pub. 843, NAS-NRC, pp. 143-156.
    Hackler, L. R. (1977). Cereal Chem. *54*, 984-995.
    Torun, B. (1978). *In* "Soy Protein: Impact on Human Nu-
trition" (this conference).
    Viteri, F. E. and Alvarado, J. (1971). *In* "Recursos,
proteinicos en America Latina," (M. Behar and R. Bressani,
eds.) pp. 53-77. Institute of Nutrition of Central America
and Panama, Guatemala City, Guatemala.

[2]Following discussions at this conference, Dr. Scrimshaw
has suggested that the recommendation that the RPV assay be
used for protein quality evaluation instead of the PER assay
(as proposed for inclusion in the Revised NRC Publication 1100
on Evaluation of Protein Quality) may require some modification to emphasize that no rat assay procedure is fully satisfactory for predicting the value of a protein in practical human nutrition. Drs. Scrimshaw and Harper suggest the use of
amino acid scores when it is not possible to have data from
metabolic studies in human subjects.

Viteri, F. E. and Bressani, R. (1972). Bull. WHO *46*, pp. 827–843.

Womack, M. (1978). Personal communication.

Young, V. R.; Rand, W. M.; and Scrimshaw, N. S. (1977). Cereal Chem. *54*, 929–948.

DISCUSSION AND COMMENTS FOLLOWING PRESENTATIONS
BY PANEL MEMBERS

Hopkins: Before I open the panel for general comments, I
think it would be appropriate to ask Dr. Satterlee to comment
on his research on *in vitro* assays for protein quality.
There has been considerable discussion on *in vitro* assays for
predicting digestibility and protein quality. Dr. Satterlee
has had some experience in this regard.

Satterlee: I would like to keep my comments very short.
We have had discussions during the last two days, and I think
it has come up several times today, that we may want to look
into methods of measuring protein quality that utilize chemical score. We at Nebraska in the Food Science Department have
been working on a procedure that utilizes the essential amino
acid composition along with the *in vitro* protein digestibility of food proteins. We compare these two properties of a
food protein to the same properties found in ANRC casein, and
from that predict the protein quality. This is a unique procedure in that it differs from the basic chemical score in
that we are looking at all of the essential amino acids. But
we almost always find that four of the essential amino acids
are very critical: lysine, methionine, threonine and tryptophan. Then by the use of a computer model, which will determine the degree each essential amino acid is limiting, and use
in vitro digestibility to get an estimate on how much of EAA
may be available, we are able to predict protein quality. The
model has been built around 52 different foods, ranging in
corrected PER values from as low as 0.69 to as high as 3.22.
Apparent digestibility values range from 65% to 92%. We can
predict a PER value within about 72 hours. Within the short
time I have here, I cannot get into detail on the calculation
procedure. This procedure will be reported in detail at the
IFT Meeting to be held in Dallas, Texas in two weeks.
We are using the PER as the basis for our prediction; but
we have no love for the PER assay, as I find many people do
not. But we find that there is a real need at this time to be
able to help our food industry. When we received an NSF grant

341

to do this work several years ago, the main concern was that
the food industry, through their quality control laboratory,
was unable to keep track of protein quality via ingredient
changes or processing changes that may occur. We felt it was
important to relate to a protein quality value that they were
using, and that was the PER. Secondly, regulatory people
might also use the new procedure, rather than PER, for their
regulatory purposes.

Our ultimate goal, of course, is not to predict PER, in
fact not to predict any quality of the rat, but to move toward
trying to figure out what the human would do in response to a
protein source. We have begun moving in this direction by
working with several researchers, getting information and data
from them, as well as developing some human data by ourselves.
By trying to understand what is involved in the human determination of protein quality, we are now looking at digestibility
- that single factor, and seeing how human enzymes, *in vitro*,
actually digest a protein. We have seen in some cases that
human enzymes differ from those of rats in their ability to
digest protein.

Our goal, as I have said, is to develop a rapid assay. We
are not trying to replace any bioassay that is available. We
are mainly trying to help the food industry come up with something that they can use on a day-to-day basis or an hour-to-
hour basis to figure out where they are on protein quality.

Hopkins: Thank you very much Lowell. I would like to
open this panel to discussion and comments from the floor or
to questions that you would like to ask the panel.

Burnette: I wanted to make two comments. One is relative
to what was just said about the problems that the food industry has with protein quality assays and with one of the traps
that we all fall into. We talk about the protein quality
tests, we talk *ad nauseam* about the different protein quality tests, all of which were designed for testing protein.
None of these have much impact upon the tests on actual food
systems required by the food industry, especially if they assume the legal liability of nutrition labeling. For that reason, at GMA we now are finishing up a collaborative test on
PER; not because we love the PER either, but because it's in
the law. The collaborators in the test are ten industry laboratories, two private testing laboratories, and Food and Drug
Administration's laboratories. A good portion of this test
has to do with sample preparation, sample handling and trying
to get an estimate on the protein quality of actual foodstuffs, as they are found, with their levels of fat, salt,
moisture or however they are presented to the public.

The second comment that I wanted to make has to do with Dr. Vanderveen's comment on the need for compliance. I am certainly not going to come up here and say that there is a need to abandon all compliance, but I also don't think that there is any justification in compliance for compliance sake. There are such things as the qualities of proteins. There are not such things as requirements for protein. We still find ourselves, in scientific meetings, falling into the trap of talking about protein requirements. There never has been a requirement for protein and there never will be. Yet that's the way we think of it, and that's the way we present the information to the consumer.

As a matter of fact, in Dr. Carroll's talk this morning, he presented an interesting point. I couldn't find an appropriate time to discuss it then, so I saved it for now. Plasma cholesterol of rabbits fed animal sources of protein was higher than those fed sources of plant protein. If you look within either one of those two groups there was an inverse relationship between protein quality and plasma cholesterol. Yet, in some later slides, he showed that when you mix plant and animal protein so that you lower the protein quality of the mixture, you also lower the plasma cholesterol. I don't know what it means. I don't claim to even understand it, but it does indicate that there may perhaps be qualities of protein that are even different from the amino acid levels that are involved in growth or maintenance of animals or humans.

In talking about qualities of proteins we may be talking about providing the consumer with two important things. One is information. The other is, for lack of any better term, economic protection, which Dr. Kies brought up in discussing the possibility that a new process may reduce the use of a historically used and known protein source. If, in fact, we are talking of two cases, we have to look then at two totally separate systems. We can write into regulation economic protection or biological protection. We can't very easily convey this information to consumers without making them all nutritionists. And, in conveying information to consumers, I am concerned that we are all tangled up in problems which are not very relevant. Perhaps the ultimate in indicating how we can go wrong with protein quality tests was shown in the last decade by Dr. Ken King's work down in Haiti, in which he used about as simple a mixture as possible - a simple bean and corn mixture from the native food supply - and raised perfectly healthy infants on two otherwise very low quality (regardless of the test) proteins. If he can raise infants in Haiti by feeding two such protein sources, any attempt to regulate, or to convey information to consumers based on any assays which would be done on either one of those proteins individually (or

all the protein sources that are in the mixed diet which is
prevalent in our society) is not conveying useful information
or protection.

I think we are at the point where, both within the National Academy of Sciences and within our regulatory bodies, we
have to think about moving away from talking about protein requirements and moving toward as simple a system as we can.
This system probably will be, and I hope we don't end up with
more, total protein and methionine and lysine levels. It has
been pointed out before that these are perhaps the most important.

We should move some of our research effort away from looking at different assay methods and toward one of the problems
that Lowell just brought up, that is a chemical digestibility
test or availability tests. We can't say we are going to use
a chemical score and have a quicker assay for protein, and
then run a rat assay to get digestibility, because then we are
back doing an animal assay. Very little work has been done,
except the work that Lowell's group has done, and a lot more
needs to be done before regulatory agencies would buy a standardized *in vitro* methodology for assessing availability,
digestibility, and other factors correlated with human tests.
Such a value could be used with two or three simple amino acids scores to relay concrete information to the consumer, the
same way we do with pyridoxine or calcium.

Pellett: I think it's rather nice to look at what happened at the international agencies. If you look at the nam-
ing of the group that was concerned with protein, it was the
Protein Advisory Group, it became the Protein Calorie Advisory
Group, and now it's the Nutritional Advisory Committee.
That's rather relevant to what has happened in this whole
field of protein. I think it is remarkable that we now have a
considerable degree of unanimity; and there is much agreement
that while protein quality may have some importance in some
circumstances, it is something that we have overestimated in
importance tremendously over the years. It is overestimated
because it has been scientifically interesting and because it
has made nice thesis material. The whole concept has relatively minor practical importance for human nutrition. We
consume diets, not proteins. As such, if these are consumed
at an adequate level of intake, not too many diets are very
different in their protein value. When we come to digestibility, I agree we need to know more, but in practice most of the
proteins that we eat in most of our diets have a high level of
digestibility. Therefore, if we measure the amino acid level
and then use some estimate of absorption, we are pretty close
to what we need to know. The fact that we are pretty close is

very important. I think we are arguing too much about the
relevance of protein quality. We accept the fact that we can
put down the iron content as iron content while we all know
there may only be 10% absorption. We don't argue about that,
but we worry tremendously about the situation of measuring
protein quality.

Altschul: What I have been hearing is a discussion of
possible solutions, but I have not really heard the clearly
defined problem. Let me try two different kinds of approaches. I think that if I were given a can of something
that is unknown and it had, let us say, 50% protein, and someone said, "Would you devote two years of your life to determining whether it is a good source of protein?", much of what
you people have said is relevant. That is, how would you go
about determining protein quality? What should be the steps
that you would take? But suppose I gave you a can of, let us
say, soy isolate, which is not a new protein. It has been
around for quite a while and we have heard just a little bit
of the enormous amount of information available in human experiments on this material. Then I wonder if it is not proper
to ask, "When do you drop the bioassays and come up, instead,
with a set of chemical numbers that will define this product
accurately enough to be able to satisfy the consumer and to
satisfy the regulatory agencies?" I'm not sure we even have
to go at this point to Dr. Satterlee. It may be that his is
for intermediate cases. The point that I am making is that
one of the characteristics that distinguishes plant proteins
from animal proteins is that animal proteins generally are
more alike and less variable. You can, of course, in nonfat
dried milk have an enormous range of protein quality by the
way you dry the milk, but that is an exception that comes
about through processing. In general, animal proteins are not
quite so variable, but plant proteins are enormously variable,
depending both on their composition and how they are handled;
and I think that it is almost an impossibility, to begin with,
to try to develop a uniform method of dealing with them as a
group. Maybe the way to do it is to take one at a time. Let
us say, take soy isolate and ask, "What is the simplest and
most reliable way, both for the consumer and the regulatory
agency, to define specifications of a suitable material?" If
I were writing specifications on cottonseed, I don't care what
you do, I would insist on a maximum gossypol value. I would
want to know the gossypol content, because I know that this is
present in cottonseed. I suppose if I were doing peanuts, I
would like to know the aflatoxin value, regardless of everything else that is said. So it seems to me that unless we are
very specific about what we are talking about and refuse to

generalize beyond what we can, we are going to keep going around in circles by talking in generalities.

Bodwell: Dr. Altschul, surely one of the things that's given research on protein quality a breath of life is nutritional labeling. Protein is the only nutrient for which a quality factor is considered in nutritional labeling; and when FDA put in breakpoints in PER of 2.5 and 0.5, it became an economic necessity to start looking at protein quality. From this viewpoint, several of us here have argued that surely there is a very wide range in protein quality in proteins having a PER between 0.5 and 2.5. In the American diet, we are eating a mixed protein diet. The breakpoints don't make sense. But, as I have said before, the regulatory agencies came up with some compromises under pressure when they came up with a labeling system. I think the argument here is that if we move away from the present labeling system which is being used, what do we use? I think protein quality and the assay methods used become extremely important. It's a matter of economics. I also very strongly feel that if we move to another approach, the new approach should have been thoroughly validated by human studies before being implemented.

Vanderveen: Well, I can see some of the rationale for what Dr. Altschul is indicating. As a regulator I would like you to consider our problems in attempting to not overregulate. What I think I heard today was the concept of getting away from specifications (i.e., PER type requirements) and control each vegetable product individually on its perceived merits. That is, soy can be used in certain products such as infant formulas, whereas other vegetable proteins cannot. I also heard that we, the regulators, should let the industry and society make the decisions about where vegetable proteins can or cannot be used in the marketplace. I subscribe to some degree to the philosophy of the latter suggestion; however, we must have some basic rules where harm might occur. For example, we cannot permit foods which are used, or could be used, as the sole source of nutrients, such as infant formulas, to be made from proteins having the quality of unfortified wheat gluten. In order to deal with the protein quality issue some means of assessing protein quality must be used. As I indicated earlier we understand that current official methodology results in underestimates of the value of vegetable proteins. The important point is for the scientific community to develop and agree on better methodology which can be used to replace PER. We don't think that regulation of the use of vegetable proteins on a product-by-product basis is the answer.

Torun: I would like to make two comments. My first comment is somewhat related to the one made by Dr. Altschul. We have been talking and hearing about protein quality, but I have not heard many things said about the fact that protein quality has to be evaluated in the context of food quality or total diet qualities. I think we should consider that there are also other factors in foods that might affect protein quality. I agree that it is not practical to try to evaluate all possible factors, but just as several comments were made on the importance of defining very clearly the type of soybean protein that was used to test the availability of trace minerals, it also is important to standardize the conditions under which the protein quality should be evaluated. I think this is of particular importance where the other components of the diet might affect the bioavailability, not of a pure protein, but of the proteins present in the whole diet being used. In short, we should really be thinking about *diet* quality instead of *protein* quality.

The other comment that I want to make is about something I heard once today and once yesterday, questioning the "morality" of using human assays. I have not heard, however, any questioning of the "morality" of using animal assays for applications in humans. I agree that when we carry out human assays, we have to set very strict guidelines to protect the experimental subjects. There should be no doubts here about what is and what is not moral. For example, if the well-being or the health of a human subject is jeopardized, that definitely is not moral. But if we follow all possible safety procedures, I wonder if sometimes it is not "immoral" to use animal assays, extrapolate the results to humans and risk preventing good nutrition from reaching humans simply because the food tested was not good enough for the rat or for the monkey. On the other hand, we must consider that in extrapolating from one species to another we can err on either the "plus" or the "minus" side. It is conceivable that an animal assay might result in the erroneous interpretation that a food which is good for the animals is also good for humans. If this conclusion is false, the "experimental" group that will prove its falsehood will not be a group of human subjects studied under strictly controlled metabolic ward or laboratory conditions, but a whole population.

Van Stratum: I will not try to add to the confusion by giving you some European considerations. In thinking it all over, it is quite clear that a lot of people need methods to determine the quality of protein sources.

I do not only mean quality control of protein as such, I also mean quality of protein sources. Different people have different aims. I will describe a few of these.

There is some concern, first, about the quality of the
protein supply for specific population groups. Secondly, introduction of novel protein sources or novel products in man's
food raises questions. And thirdly, there is a need for quality control, both for industry and for the consumers, including regulatory authorities. For whatever reason people are
concerned about protein quality, all have only one word in
mind, that is PER. They think of PER as a standardized test
like the octane value of gasoline. People seem to think that
a human machine does not operate properly if it gets some
poorer quality protein. For all these useful qualifications
of quantifications of protein sources, in Europe we think PER
is not an adequate measurement. Therefore, this would make
the use of animals for PER determinations questionable, as
Dr. Torun said.

PER is not good for consideration of protein intake of
populations as a whole. Enough has been said of the quality
of single proteins. No problems. We eat mixed diets, we eat
variety, we eat actually in excess of minimal needs as soon as
we can meet our energy requirements and afford a choice of
foods.

The second point concerns the intake of novel proteins or
novel products to be introduced in human diets. Well, I admit
consideration must be given to this. The effect of consumption of these products has to be tested not only for the protein nutrition aspects as such, but for all other aspects of
human physiology.

The third aspect just concerns quality control. I think
using PER is the poorest way to spend money, animals and man-
power. I agree with Dr. Altschul, that we are challenged to
look for other methods tailored to each product, and to look
for the weak points in technology apart from protein only. I
think that it is foolish to try to satisfy everybody, with every kind of product, by using one single type of measurement.

If, nevertheless - what I consider to be the worst situation - some law or regulation forces us to do PER determinations, we have to accept that grudgingly. But at the moment,
I want to reserve the right to appeal to a higher court if
this PER fails to demonstrate the true position of a protein
source. In both the interest of the producer and of the consumer, I must be able to choose another method, be it an animal test like slope ratio or NPU or even human testing. Finally, there is need for continuous discussions with all people who are thinking PER, including scientists and even some
nutritionists.

Hopkins: Thank you. We've got just a few minutes left
and I'd like to direct a little bit of questioning toward the

suggestions that two of the panel members made. Dr. Steinke talked a little bit about a new method of measuring protein quality. I would like to get some reaction to that. Ricardo Bressani suggested the short nitrogen balance; I would like a little reaction to that.

McLaughlan: I would like to comment on Fred's suggestion. Basically, it is just a form of chemical score, tied in with availability, right?

Steinke: That's correct. It's essentially using the human scoring pattern.

McLaughlan: I think that's the direction you should be going, all right. Right at the moment there is still the problem of determining available amino acids; and maybe when that's solved, we won't even have to bother with an animal assay at all. I don't know. But certainly that is the direction we should be going. If we were starting right now to devise a method to measure protein quality, we probably wouldn't even think of an animal assay. We would try directly to develop the chemical score method.

Steinke: It is difficult for me to go back, not having been in the area for 50 years, know exactly how we arrive at a standard of 2.5 for a PER assay to begin with. I'm sure there was some logic at some point, but it's a little foggy at this point.

McLaughlan: I think the 2.5 was derived from rats that were not all that healthy when casein was used to give a PER of 2.5. Now, it's usually more like 3.0.

Steinke: But I think the point is that it was never really correlated to human requirements.

Bodwell: I would like to start out by saying that the RPV, which according to a lot of people is theoretically a sounder rat assay than PER, may not be any better when it comes to predicting the protein nutritional value for the human than a theoretically inferior assay - the PER.
I would like to comment a little on the short-term nitrogen balance method that Dr. Bressani discussed. Theoretically, I think the approach can be criticized very strongly and Dr. Bressani has alluded to this; it is very comparable to the long-term balance assay, studies which take 30, 40 or 50 days. Now, the first criticism; the long 3- or 4-day non-protein adaptation period is certainly a period of stress. Who knows

what happens in various body tissues or constituents such as
liver enzymes during that period. Then, when his subjects are
moved up to 0.2 g protein/kg body weight (which even MIT has
quit using because it has given a skewed point on their regression line, just like the NPU), one is exaggerating the
protein quality at these low intake levels. At marked negative nitrogen balance one is tilting the line. So really,
there are only two valid points on the regression line, at 0.4
and 0.6 g protein intake/kg/day. However, although you can
knock the approach all you like, the actual results are very
acceptable when compared to long-term assays.

I would make one further comment, though, after seeing
your data, Dr. Bressani. You really are biasing the plot by
entering the zero intake values. Just like the old-fashioned
slope ratio with rats, your slope would be very different if
you left off that zero intake point. The slope would be tilt-
ed and then you wouldn't be getting agreement with the long-
term assay. So actually, what seems to be a negative factor
theoretically; the 3-day non-protein period is actually making
the slopes comparable to those obtained in the long-term conventional approach.

Bressani: The multiple point assay is a new term being
used today for methodology developed some time ago, and is to
be found in the old literature of protein quality evaluation.
The mathematical analysis of nitrogen balance data from variable intakes establishing the relationship between nitrogen
intake and nitrogen retention clearly states that when N intake is zero, nitrogen retention is, and must be, equal to total endogenous N excretion from feces and urine. This, in our
opinion, is the true biological point in the equation, which
must be used in the mathematical expression of the protein
quality as determined by the nitrogen balance index. We
therefore use this value in our calculations. Furthermore, we
consider the assay to be valid only if the mathematically derived value of "a" in the equation $Y = a + b X$ is close to the
true total nitrogen excretion from feeding a nitrogen-free diet. The actual estimation of total N excretion could be eliminated from the assay provided enough data were available for
this value. Finally, the protein feeding which causes some
problems in establishing a linear relationship between protein
intake and nitrogen balance is the first level after the
nitrogen-free diet feeding. Furthermore, this value is responsible for overestimating protein quality in most situations. In our experience with dogs, we eliminate this problem
by feeding small increments of protein rather than going from
0 to 0.2, for example. This is an area which requires some
research, and there are ways to get around this problem.

Kies: I have two comments that I wanted to make. First of all, on Fred Steinke's proposal, although we feel the digestibility is a very important point to consider in alternative determination of protein qualities, I think that there is also a danger in going overboard on digestibility. As stated earlier by Dr. Scrimshaw, one must remember that purified amino acid diets are very poorly utilized by humans even though they are completely digestible; they are really predigested foods. There is such a thing as things being too digestible.

The other thing that I've been wondering about in this short bioassay method - what do you do about fecal assays? These are hard to do; and on a 2-day assay, how do you separate feces from assay from non-assay days? We have problems on 4-day periods.

Bressani: When we started developing the short time nitrogen balance index assay we were doing twenty-four hour collection of feces and urine. It was found that it was not always possible to get representative fecal collections. Because of these it was decided to shift to a 2-day collection period, which we feel improved the assay. Because of this problem now we choose the experimental subjects on the basis of the frequency with which they defecate. Furthermore, we use markers. Even with these modifications quantitative fecal collections are not as efficient as we would like them to be. We have also studied the possibility of pooling all fecal material during the 6-day balance period and distributing the fecal nitrogen proportionally to the different levels of protein fed. However, up to the present time we have not been completely successful because protein digestibility at levels of intake of 0.2 or 0.3 g/kg body wt/day are not exactly the same as when protein intake is increased with a 0.6 or 0.7. Probably the approach would be to distribute fecal nitrogen output on the basis of daily dry matter excretion.

Bodwell: One of my comments to my human subjects is that we actually are getting very biased results any time we run a study with human subjects, because obviously a normal person, a typical person, would not be a volunteer in a human study. So the sample is biased. If Ricardo is screening his subjects to obtain those which have a bowel movement every 2 days or less, I think that is an even more biased sample. Anyone who is running human studies starts in with the study and finds out 5 days later that one or more subjects have had no fecal collection. You immediately think there is a constipation problem, which can be a serious problem. Then you find out from the subject that for that subject it's very normal to have a bowel movement only once every 5 days or so. It plays

hell with your markers and in a short-term approach, obviously
that subject must be eliminated. People are different and
someone that has 5 collections in 5 days is different from
someone who has 1 collection in 5 days. I don't know what im-
plication this has with reference to a short-term assay.

   Zezulka: I wanted to address Dr. Steinke's suggestions on
digestibility. I think if you are going to try to correct the
amino acid requirements of rats to humans, I can understand
doing that; I'm not sure, though, that looking at the digestibility in terms of PER is going to help, because there again
you are still looking at the rat digestibility. I think
that's the question we've all been addressing, and it might be
just as good to look at some *in vitro* methods for determining digestibility and really not pay a whole lot of attention
to PER at all. Maybe you want to comment on that.

   Steinke: I wasn't really specifying any specific animal
for determining the digestibility. There are a number of
modes. Obviously, the easiest for just a routine assay would
be an *in vitro* assay using enzymatic or microbiological
methods.

   Zezulka: I thought that you were saying after you did
that, then you determined PER and readjusted for known differences in amino acid requirements of rats versus man.

   Steinke: That was the second alternative proposal; if it
is absolutely necessary in order to obtain consensus of the
nutrition community to run protein quality on animal basis. I
suggested using a specific amino acid response in the rat, but
make it a different response for each amino acid based on the
rat's requirements relative to humans. There are two separate
proposals involved. Probably the pig is a better model for
digestibility than the rat. We have run a number of those and
they correlate very well with the human digestibility.

   Zezulka: Thank you. A second problem comes to mind probably because we've done our nitrogen balance studies with soy
a little bit differently than Dr. Bressani or Dr. Kies. In
looking at amino acid requirements I really think that we
should be looking at amino acid requirements per se and not
with the concomitant problem of total nitrogen requirements.
I'll agree with Dr. Kies that it does increase the differences
that you can note between proteins if you are feeding sub-
minimal requirement levels of nitrogen. You are really talking about two deficiencies superimposed on each other. In
looking at what happens during periods of negative nitrogen
balance caused by sub-minimal nitrogen intake in changes on

readily available amino acid pools, the actual amino acid requirements may be different during that circumstance than during the circumstance of nitrogen equilibrium. That's one of
the reasons that we have attempted in our nitrogen balance
studies with soy to try to maintain a nitrogen level that certainly should be adequate, and then look at the protein quality as far as amino acid content is concerned at the same time.
I wanted to bring out the point that there is another way of
looking at amino acid requirements from a particular protein.

     Pellett: Be careful about jumping on that digestibility
business. You have to look at the whole diet again. You
can't talk only about the protein.
     Casein in a diet which includes a high amount of fiber has
a different digestibility than a diet that includes casein all
by itself.

     Samonds: If mine are the last comments, I wish they were
more profound. I do have two short comments. One is to mild-
ly chide a few of our speakers. Very few measures of protein
quality that we have seen here today, whether they have been
slopes by Dr. Bressani or ratios by Dr. Bodwell, included
standard errors of these estimates. Every estimate of protein
quality, whether it be PER or NPU or NPR or biological value,
every one of those has a standard error. The correlation coefficients of the regression equations are *not* adequate for
us to decide whether there are really differences between proteins. If we are looking for differences in protein quality,
this implies that we are looking for statistically significant
differences. The reality is sometimes quite discouraging when
we are unable to show differences in quality between proteins
which are commonly considered to differ in quality, as was the
case in the collaborative assay in which I and several of you
participated.
     My second comment is with regards to the frequent criti-
cism I hear about the cost of the RPV assay. I think you all
have to remember this methodology was developed at a relatively affluent university with NIH grant money and there was
really no incentive to make Slope Ratio or Relative Protein
Value cost effective. We chose 6 animals per group and 3 or 4
levels of each protein because it sounded like a good idea and
because it usually gave considerable precision. Nobody that I

know has taken that assay and attempted to cut down on the
number of animals or to find the proper levels of protein intake to get the maximum amount of information, or the maximum
statistical reliability, at minimum cost. We always are going
to make a trade-off between our ability to differentiate between protein and the number of levels, the number of rats,
duration, cost, etc.; so, I think that's something someone
should try...to see how much one can cut out of the assay and
still get statistically significant results.

SUMMARY OF WORKSHOP: HOW SHOULD PROTEIN QUALITY BE DETERMINED

D. T. Hopkins

The question of the practical importance of measuring protein quality for human nutrition was discussed following the panel presentations. Several of the participants suggested that in the United States where protein consumption is generally more than adequate, the measurement of protein quality assumes lesser importance. One participant suggested that the quality of protein should only be measured in human subjects at levels of consumption of 40 grams per day. If quality of a food protein is sufficient at that level of intake in adults, then there need be no concern that the protein quality is compatible with nutritional well-being. It was pointed out that although this may be the case in industrialized countries, in developing countries a somewhat different situation may exist. In countries such as those in Latin America, on some days people will consume sufficient protein to meet their requirement, and on other days they will not receive enough protein because they will consume only grain and no protein supplement such as beans or meat. For example, some days people may eat only corn and other days they may eat a mixture of corn and beans. In this type of situation a measurement of the quality of grain protein or a mixture of the grain and legume protein is of interest. The ability to rank the protein quality of mixtures of grains and protein sources is also important.

Concern was expressed about the digestibility of protein by humans and the effect on protein quality. Digestibility of proteins by humans may not vary appreciably; but, on the contrary, a range of 65% to 90% was considered a possibility, depending upon the protein source. The group concluded that ordinarily, digestibility of diets by humans is high, but the cases in which it is not are of valid concern.

Models for digestibility of protein by humans were discussed. Rats may not be a good model for humans; the data do not seem to be available. Data on digestibility of pro-

teins in both humans and experimental animals should be
tabulated.

The subject of amino acid patterns for scoring proteins
for human needs was discussed. Presently there are two patterns, those of the Food and Nutrition Board and those of the
FAO/WHO as outlined in the panel discussion by Steinke (loc
cit). A consensus is needed on the appropriate pattern to be
used in protein scoring. An effort should be made to concentrate further research on amino acid requirements of humans.

It was concluded that a combination of digestibility and
amino acid scores is a viable procedure for measuring protein
quality in humans. However, protein quality values obtained
in this manner will have to be validated from time to time
with human assays. There was considerable discussion on the
design of protein quality studies with humans and the interpretation of the results. In Bressani's panel discussion some
of these problems were outlined, particularly the question of
linearity or non-linearity of nitrogen balance values at different levels of protein feeding, and the meaning of the
extrapolation of nitrogen balance data to the point of zero
intake. The workshop group was concerned about the acceptable
degree of variability in human assays and how to deal with
this variability either through experimental design or
increased number of subjects.

It was concluded that a meeting should be held in the
future to explore further possibility of using some estimate
of digestibility and protein scores to predict the quality of
proteins for human nutrition. The meeting should be held
under the auspices of some national or international
organization. At such a meeting the following points need to
be considered:
1)   A comparison of protein quality values obtained from human
feeding studies with those predicted by utilizing a combination of amino acid scores and digestibility values.
2)   A summary of protein digestibility by humans and by
experimental animals.
3)   Appropriate amino acid pattern requirements to be used for
computing protein scores.
4)   The statistical design, analysis and interpretation of
human bioassays of protein quality.

REFERENCES

Bressani, R. (1978). A Critical Summary of a Short Term Nitrogen Balance Index to Measure Protein Quality in Adult Human Subjects. This volume.

Steinke, F. (1978). Measuring Protein Quality of Foods. This volume.

WORLDWIDE REGULATORY PERSPECTIVE
OF PLANT PROTEINS IN FOODS

Leonard H. Roberts

INTRODUCTION

It is obviously fitting that this Conference give attention to the regulatory developments regarding plant proteins -
if only because these developments are currently well under
way on a global scale.

In the United Kingdom the Ministry of Agriculture, Fish-
eries and Food is about to issue a legislative framework for
the usage, labeling and nutrition for plant proteins. The
recommendations by the U.K. Food Standards Committee, in its
1974 Report on Novel Protein Foods, will undoubtedly have a
significant influence upon the ultimate complexion of this
legislation (Food Standards Com., U.K., 1974).

On April 15, 1975, the Canadians issued one of the most
elaborate set of regulatory provisions governing the usage of
vegetable protein in meat and poultry products (Canada Food
and Drug Reg., 1975).

On August 27, 1975, the French Government issued its first
provisions governing the definitions and labeling of vegetable
proteins (French Minister of Agr., 1975).

And the list goes on. Norway, The Netherlands, Italy,
Venezuela, and Japan are all in the process of elaborating
national legislation governing the usage, labeling and nutrition for vegetable proteins. Spain has recently joined the
ranks of others in the global challenge of developing national
regulatory policies on the same.

Closer to home we find one of the most dynamic efforts occurring in the development of both direct and ancillary food
law policies on vegetable protein usage.

Regulatory development regarding the use of vegetable proteins has been a matter of continuous review by the United
States Food and Drug Administration and by the United States

Department of Agriculture. On June 14, 1974, FDA issued its
proposed Common or Usual Name for Plant Protein regulation
(Fed. Reg., 1974). Among other things, this proposed regulation purports to establish both labeling and nutritional
guidelines for the use of plant proteins in meat, seafood,
poultry, eggs, and cheese food products.

Independent of the issue of vegetable protein, there are
many food law issues that are currently undergoing vigorous
debate in the United States, that although ancillary in nature, will directly impact upon the evolution of the future
legislative framework for vegetable proteins. Among these ancillary developments include the debate on FDA's safe and
suitable optional ingredient concept, nutritional labeling
regulations, the imitation labeling policy, policies governing
nutrient fortification, and probably most important, the present debates and deliberations that are now addressing the
efficacy of establishing strict recipe-type standards of identities for processed foods. Historically, these recipe standards have either precluded or greatly limited the participation of vegetable protein in traditional products.

The economical, nutritional, and advanced technological
development of vegetable proteins have truly created a challenge to ingenuity in the construction of new food laws and
regulations throughout the world. This regulatory challenge
has already gone beyond any of the previously mentioned national efforts. Several international regulatory bodies and
functionaries have now given this subject matter a high priority.

In 1974 the Commission of the European Communities in
Brussels established a special study group "to consider the
scientific, technological and health aspects of the commer-
cialization of foodstuffs containing vegetable protein." The
final report by the Study Group on Vegetable Protein in Foodstuffs for Human Consumption will be a principal consideration
by the Commission in its eventual elaboration of legislation
on this subject matter. In December of 1977, this special
Study Group issued its draft report for comments (Com. of the
E.C., 1977). We are told that the final Study Group report is
in its final stages of completion and will be shortly submitted to the Commission.

The highlight of this global regulatory inertia on vegetable protein took place in Rome less than one month ago. It
was in Rome, on April 26th, and during the Twelfth Session of
the Codex Alimentarius Commission, that a new independent subsidiary body within the Codex framework was officially established, namely, the Codex Committee on Plant Proteins. This
new Codex body was given the broad mandate as articulated and
recommended by the report of the Secretariat of the FAO/WHO

Food Standards Programs (Codex Alimentarius Com., 1978). The recommended mandate, which was subsequently supported by the member country delegations, gave the new Committee the charge to elaborate definitions and worldwide standards for vegetable protein products. The Committee's mandate also includes a charge to elaborate guidelines on the utilization of vegetable protein, on nutritional requirements and safety, on labeling and on any other aspect as the Committee may deem appropriate.

A detailed analysis of all or even any one of the mentioned regulatory developments would be far beyond both the scope and time constraints of this presentation. However, a few observations and comments from a review of this global regulatory development is surely warranted.

FOOD CLASS ORIENTATION

The orientation of national regulatory developments, proposals, and ideas has been towards the use of vegetable protein in meat products. Such orientation would seemingly be rather narrow, and in fact, it is. Meat products are only one of the many vehicles through which vegetable proteins find their way into the world food supply system. Vegetable proteins are also used in the baking industry, the dairy industry, by the dietary food industry, and by the baby food industry. Furthermore, in most of the developing countries of the world, vegetable proteins are being used in simple mixtures to improve the nutritional quality of staple foods. While processed meat products contribute substantially to the food supply systems in developed economies, they constitute only a minor component in the food supply of developing areas.

REGULATORY APPROACH

The traditional regulatory approach for vegetable protein legislation has been "vertical" in nature. Vertical legislation is a term that has been used to denote legislative action which effects changes in the compositional laws of foods by the process of amending individual food standards. For example, legislative reformation to increase the allowable usage level of vegetable protein in foods may be effected by amending a selected number of individual food standards.

In the past, the vertical method of legislating for vegetable protein has been the almost exclusive approach taken in
the U.S., particularly by the U.S. Department of Agriculture.
This is now changing in the U.S. and with the rest of the
world. The approach in legislating for plant proteins has now
become more "horizontal" in nature.

The horizontal regulatory approach accepts as its theoretical basis a given set of national food standards, and attempts to allow for vegetable protein usage by labeling and
nutritional guidelines. In other words, horizontal vegetable
protein legislation would not deal with standardized products
in particular, but would with all foods (or classes of foods)
in general. Hence, its orientation is generally far more en-
compassing in scope and not necessarily limited to meat products. Labeling guidelines which establish rules for adding
vegetable proteins to otherwise standardized food products
(without disturbing their respective standards) is an example
of the horizontal approach. In addition to labeling guidelines, hygienic guidelines on the vegetable protein itself (as
a food ingredient) is a typical component of this type of regulatory approach. Also, in certain countries, nutritional
guidelines have been considered a facet of the horizontal
approach, particularly when a portion of an otherwise required
level of traditional characterizing protein ingredient of a
standardized food is substituted by vegetable protein.

Upon analysis, one would recognize that the horizontal
approach is the underlying basis and direction of FDA's proposed plant protein regulation. The horizontal approach is
also an approach which is conducive to international harmonization of divergent national legislation, since it need not
deal with individual national food standards. Indeed, it is
the horizontal approach that is being used by the Commission
of the European Communities, and may be the only logical and
feasible approach that will be undertaken by the Codex effort.

DIVERSIFICATION OF LEGISLATION

The purpose of this paper is to merely give you a general
perspective of the extent and nature of the worldwide regulatory direction governing vegetable protein. If I had to give
you this perspective in one word, it would have to be "diver-
sified." This diversification has and will unfortunately con-

tinue to lead to serious impediments to international trade. It is against this background of international diversification and the resulting trade impediments that the industry and many governments are supporting and encouraging the active involvement by the Codex Alimentarius Commission in its harmonization efforts with vegetable protein legislation. Codex certainly represents the unique international organ that can undertake the task of achieving international harmonization with the existing divergency in national legislation.

The two principal areas where there is a substantial divergence of notions and approaches concern:
1) The food composition and labeling law structure which determines the entry of vegetable protein into a national food system; and
2) The *scope* and *nature* of nutritional guidelines that are triggered by this entry mechanism (e.g., vitamin, mineral, and biological quality requirements).

More interesting and relevant for this Conference is the international divergency in both the scope and nature of the application of nutritional quality guidelines. We will therefore limit our discussion to this area.

Much regulatory attention throughout the world has been focused on the issue of whether or not vegetable proteins, particularly soy protein, should be fortified with vitamins and minerals and/or amino acids, namely methionine. The issue is far from being simple and has been the center of a considerable amount of global debate and controversy over the last few years. The two principal regulatory issues have been "when?" and "what?". *When* should vegetable proteins be fortified and *what* should be the fortification requirements?

An attempt will be made to identify as objectively and as briefly as possible the varying governmental positions and trends on these two questions. It is hoped that this discussion will provide a better understanding and insight into the status and direction of the regulatory and legal debates that are presently occurring regarding the issue of nutrient fortification of soy protein.

The issue of "when" is not to leave the presumption that all countries are in agreement that soy protein should in fact be fortified. This presumption would be incorrect since there are differing viewpoints on this basic issue.

For example, the present legislative trend and thinking in Europe appears to be to regulate traditional products containing vegetable protein as new products consisting of a mixture of traditional and more novel food protein ingredients. In other words, this European trend views the mixture of soy protein and traditional proteins from meat, poultry, dairy, etc., as products of their own identity and right. Concomitant with this European trend are the requirements for clear and informative labeling, thereby assuring from a consumer standpoint that these new product identities are maintained in the marketplace.

In Canada and the United States, however, the trend and approach appears to be somewhat different. Traditional products containing vegetable protein are considered "reformulated" traditional products, borrowing their image from these traditional products. Consequently, the regulatory requirements for nutrient fortification is the trend in order to theoretically achieve "nutritional equivalency" to "reformulated" traditional products.

The basic rationale is that the consumer expects, or theoretically should expect, a certain nutritional profile from traditional products, particularly if the composition of these products has been specified or controlled by a long established standard of identity. Therefore, any reformulation of these traditional products should not result in a deviation in this nutritional expectation, whatever it may be. Note, according to this "expectation" rationale, nutritional adequacy, or one could even argue nutritional superiority, is not the issue. The key factor in satisfying the legal equation becomes nutritional equivalency.

The dichotomy of these differing national positions on nutrient fortification became apparent during the debates at the last Codex Meeting on Processed Meat Products, held in Copenhagen in 1976. The minutes of that Session reflected that the United States government stated that "its approach would favor fortification of soy protein so that if a vegetable protein product looked like a traditional product, it should be nutritionally equivalent to the product it simulated (Codex Alimentarius Com., 1976). However, other governmental delegations, particularly from Europe, expressed the view that it would be better to let these products stand on their own merit, without fortification requirements, subject to clear and informative labeling.

Going beyond the basic issue of whether soy protein should be fortified, there has been a considerable amount of global deliberation on when, and again if at all, fortification requirements should be imposed. Hence, the issue is directed more specifically at the scope of national nutritional guideline regulations. In other words, once fortification requirements are established, when should they be applicable or imposed on vegetable protein? Here again there has been a substantial divergence of approaches and notions in resolving this issue. For example, the Canadian legislation would require that soy protein be fortified with 12 different vitamins and minerals whenever it is added to a meat or poultry prod-

uct, regardless of whether it is used as a functional ingredient (at low levels) or used as an animal protein alternate
or extender (Canada Food and Drug Reg., 1975).

Under the proposed FDA Plant Protein Regulation, it would
appear that the approach in the United States would exempt
vegetable proteins from nutrient fortification requirements
when they are used for technological purposes. Furthermore,
the particular legal basis of the proposed FDA regulation
would seem to even further narrow the scope or applicability
of the proposed nutritional guidelines. As stated in the preamble of FDA's proposed Plant Protein Regulation, the intent
of the proposed nutritional guidelines is to "identify the nutritional qualities of plant protein product extenders and replacements that will avoid the necessity of their being labeled imitation" (Fed. Reg., 1974). Thus, the legal basis for
the nutritional guideline appears to be the Imitation Labeling
laws. It would, therefore, seem to follow that the proposed
nutritional guidelines would not be applicable to vegetable
protein containing products which are *not deemed* to be imitation under the Imitation Labeling provisions of the Food,
Drug, and Cosmetic Act and its governing policies (Fed. Food,
Drug, and Cosmetic Act).

On this legal basis, it would then appear that vegetable
protein would be exempt from nutritional guidelines when they
are used for functional or technological purposes or added to
foods which are not subject to a standard of composition, or
when added to foods whose existing standards allow for a specified level of vegetable protein inclusion. By law, in each
of these situations the issue of Imitation labeling never
arises.

In most countries where the trend is to require nutrient
fortification of vegetable protein, the trend is clearly one
which reflects a pattern in terms of scope similar to that
which has been proposed by FDA (Codex Alimentarius world review of vegetable protein legislation; and the EEC Work Study
Report).

Another way of viewing this is when one finally sorts
through all the debates on the use of vegetable protein in
their various forms and multiplicities of uses; the only rational situation where nutritional equivalence would seem to
be applicable is when vegetable protein is used to replace a
portion of the traditional protein that is otherwise expected
or required by the national food law structure. When vegetable protein is used for a technological purpose, or used in
products wherein the resulting level of required traditional
protein is not specified and/or not diminished, then the imposition of nutrient fortification requirements is difficult to

justify. In each of these situations, the practical nutritional efficacy of the traditional food product has not been
altered.

It must be emphasized that this discussion on FDA's Plant
Protein Proposal is based only on a proposal and not a final
rule. It is quite possible that the scope of FDA's final
nutritional guidelines on Plant Proteins may be expanded beyond its original imitation law basis.

It also must be emphasized again that not all countries
have imposed nutrient fortification requirements on vegetable
protein. In fact, of the eighteen countries that have developed national regulatory policies or regulations on vegetable
protein, only one country currently requires nutrient fortification. That country is Canada. However, the current thinking in the United Kingdom, Norway, and the United States would
seem to favor nutritional guidelines that would necessitate
nutrient fortification of soy protein.

With respect to the nutritional guidelines themselves (the
"what" question), the international divergency is absolute.
In other words, no two countries are in agreement, both in
terms of vitamin and mineral guidelines or on guidelines concerning protein quality. These differences appear to be based
more on differing national legal principles and/or on the lack
of a uniform nutritional data base on vegetable protein rather
than based on a determined nutritional need of the national
population in question.

The differences in protein quality guidelines on vegetable
protein is of particular concern since they are compounded by
the lack of an agreed upon method for assessing it. PER, NPU
or other methods are now under active consideration by various
international governments. Furthermore, for labeling purposes
there are differences in the mode in which governments require
the reporting of the nutritive value of the protein. For example, the regulations of FDA report on protein quality, the
ones in Canada report on protein rating, others on PER, and
most recently the U.K. is considering the reporting based on a
measurement of methionine.

It is because of these current differences on protein
quality, and their potential impact on the future regulatory
destiny for soy protein, that the substance of this Conference
is both timely and critically important. New legal dimensions
are being formulated that may determine the nutritional worth
and future of soy protein. Legal theories and principles such
as the imitation food labeling laws are becoming, in essence,
the theoretical basis for protein quality guidelines on soy
protein. Consequently, the nutritional value inherent with
soy protein itself may be subordinated to its nutritional
value as measured by the food to which it is added. It there-

fore becomes paramount that continued human nutritional
research work be conducted in an effort to accurately assess
the true relationship between soy protein and other traditional proteins in the context of human dietary requirements.

CONCLUSION

Regulatory development governing soy protein is presently
in a dynamic state throughout the world. In addition to the
United States, governmental policy development on the usage of
vegetable protein has been an active venture within many national boundaries including the United Kingdom, France, Den-
mark, Holland, Norway, Spain, Italy, Germany, Japan, Venezuela, Sweden, Canada, and several Eastern European countries. Several international regulatory bodies, such as the
E.E.C. and the Codex Alimentarius Commission, have given the
subject matter a high priority. The support that the developing countries recently gave in Rome towards the establishment
of an independent Codex Committee on Plant Proteins reflects
their keen interest in this area of regulatory policy development.
This area of regulatory development has particular significance since it is at the heart of so many key issues underlying the evolution of food law development around the world.
The interrelationship of product labeling with standards of
identities, the concept of safe and suitable optional ingredients, and the imitation labeling law issue are just a sample
of basic policy considerations that are being reevaluated today. Most important, however, is an area that has definitely
underscored the entire effort to provide the regulatory vehi-
cle for vegetable proteins to enter the world food supply -
that area is *nutrition*. After all, the biological value of
any protein is the ultimate measure of its worth in the human
society.
The leadership role that the scientific community, particularly the nutrition discipline, will be playing in this complex area of food law development is unquestionably one of
great importance. This Conference and the work that is being
presented here will hopefully represent the beginning of that
which needs to be pursued in the future. I know I speak on
behalf of the global regulatory community, be they members of
the academic, governmental, industry or consumer groups, when
I say that the nutritional science leadership and direction is
a welcomed and needed input in this world regulatory endeavor.

REFERENCES

Canada, Food and Drug Regulations, Regulation Excerpts and Guidelines - Part 13, April 15, 1975.

Codex Alimentarius Commission, Report of the Ninth Session of The Codex Committee on Processed Meat Products, Alinorm 78/16, para. 77, 1976.

Commission of the European Communities, Draft Report of the Study Group on Vegetable Protein in Foodstuffs for Human Consumption - Brussels, December 1977.

Definitions and Uses of Vegetable Protein, French Minister of Agriculture, The "Circulaire" of 27 August 1975.

FAO/WHO Food Standards Program report on Vegetable Protein, Twelfth Session, Codex Alimentarius Commission, Alinorm 78/32, Rome, April 1978.

The Federal Food, Drug, and Cosemetic Act, Chapter IV, Section 403(c).

Food Standards Committee of the United Kingdom, Report on Novel Protein Foods, London, Her Majesty's Printing Office, 1974.

U.S. Food and Drug Administration, Proposed Rule for Common or Usual Name for Plant Protein Products, Federal Register, Vol. 39, No. 116, June 14, 1974.

DISCUSSION

Altschul: You talked about regulations that have to do with plant protein regulation, so I'm assuming that all plant proteins are being lumped into one category. For example, are regulations being drafted that will fit both soy and rapeseed, or are they trying to decide the legal questions on the basis of the available knowledge on each one of those materials?

Roberts: John Vanderveen may want to comment on this, but in the United States one can say that plant proteins are generally being lumped together; and as a plant protein product, one could expect the rules and guidelines will govern all plant proteins added to food products. There may be some incidental issues as to whether certain plant proteins would be permitted in a given food system. Once they are permitted, they would probably fall into this plant protein regulation. In Europe the trend is more specific. Looking at some of the international developments there seems to be a specific focus on soy proteins. On the other hand, there will be a definite interest in accommodating domestic sources of alternate food protein as they become available.

WHAT WE NEED TO KNOW ABOUT PLANT PROTEINS

Aaron M. Altschul

I want to reflect on some of what I heard; I am not trying to comment on the entire Conference; nor is this an attempt to provide a consensus. I start with the general subject of plant proteins but speak mostly about soy.

I should like to cover the following subjects:
I.   Is there a historic aspect to this conference;
II.   What are the long-term consequences of introducing soy protein into food systems;
III. Specifications of soy protein products;
IV.   The social value of introducing soy protein or plant protein in the food supply.

Let us start with the first question. We all go to conferences. Some are more important than others. Some might even be considered as marking a historic moment. Let us ask ourselves whether there are some historical aspects to this Conference. In order to try to deal with this question I ask myself, "When does a food ingredient or a food cease to be new or novel?" This is not just a semantic question but has consequences for how the food is looked upon and how it is regulated. Is soy in the class of familiar food ingredients or is it still in the vague general class of plant proteins? For some time I had no question that soy was no longer a novel food; with its long history in the orient, it, perhaps, ceased to be a novel food centuries ago. If I had any doubts, these were dispelled at this meeting. Therefore, I suggest that it is proper to propose that this meeting might mark the final transition for all of us of soy protein into the class of an accepted member of our food supply, and this has historic connotations.

This does not mean that there are no longer any problems or that acceptance is equally as well for all types of soy products. There will always be a dynamism to the concept of the information needed to define the limits of equivalence of soy-based foods for humans. It is necessary only to show that one is in the ballpark at a reasonable amount of substitution to satisfy the demands of equivalence. If, for example, soy products have protein values in a range of 0.7 to 1.3 of normally consumed protein products, this could well qualify as meeting the expectations of the consumer.

I would think that the best of all cases is when there is direct experience in experiments with humans; there is no substitute for such data. Where there is no human experience, then we have the problem of relying on the second order animal experience in animals related as directly as possible to humans in their requirements.

One of the issues in equivalence is the question of need for additional methionine. As a result of this meeting, I would expect that the issue is closer to being resolved in favor of no requirement for additional methionine. For the time being it might be that methionine supplementation of soy products for human consumption should not be mandatory unless experience in humans in certain applications warrants it; this obviously is a very important area of continued research.

The issue was raised yesterday about the need for fortification of soy products with vitamins and minerals. This raises the question of maximum flexibility for the individual versus the concept of complete equivalence. Again, it would seem to me that one ought to know how these food products are going to be used and identify the issues of equivalence that arise rather than try to develop a concept of universal equivalence for all possible applications.

But the real question is not really the question of short-term equivalence -- the real question revolves around the long-term consequences of adding soy products into the dietary of any nation. What could be expected if 10% of the meat were replaced by soy in a country like the United States? Can we foresee protein problems, or vitamin problems, or mineral problems; and how is this resolved. This question can only be resolved or approached by continuing a rather broad range of research.

There are obviously instances where long-term multi-generation animal experiments are needed. I remember in the development of the basis of support for margarine that multi-generation studies on margarine fed to rats were conducted, and they just kept going and going, and going; these continued for 40 or so generations until it was decided that no more useful purpose was served. But there is a place for certain long-term animal experiments, and there is a place for animal experiments such as the ones that were described yesterday on LAL.

the information needed to define the limits of equivalence of soy-based foods for humans. It is necessary only to show that one is in the ballpark at a reasonable amount of substitution to satisfy the demands of equivalence. If, for example, soy products have protein values in a range of .7 to 1.3 of normally consumed protein products, this could well qualify as meeting the expectations of the consumer.

I would think that the best of all cases is when there is direct experience in experiments with humans; there is no substitute for such data. Where there is no human experience, then we have the problem of relying on the second order animal experience in animals related as directly as possible to humans in their requirements.

One of the issues in equivalence is the question of need for additional methionine. As a result of this meeting, I would expect that the issue is closer to being resolved in favor of no requirement for additional methionine. For the time being it might be that methionine supplementation of soy products for human consumption should not be mandatory unless experience in humans in certain applications warrants it; this obviously is a very important area of continued research.

The issue was raised yesterday about the need for fortification of soy products with vitamins and minerals. This raises the question of maximum flexibility for the individual versus the concept of complete equivalence. Again, it would seem to me that one ought to know how these food products are going to be used and identify the issues of equivalence that arise rather than try to develop a concept of universal equivalence for all possible applications.

But the real question is not really the question of short-term equivalence -- the real question revolves around the long-term consequences of adding soy products into the dietary of any nation. What could be expected if 10% of the meat were replaced by soy in a country like the United States? Can we foresee protein problems, or vitamin problems, or mineral problems; and how is this resolved. This question can only be resolved or approached by continuing a rather broad range of research.

There are obviously instances where long-term multi-generation animal experiments are needed. I remember in the development of the basis of support for margarine that multi-generation studies on margarine fed to rats were conducted, and they just kept going and going, and going; these continued for 40 or so generations until it was decided that no more useful purpose was served. But there is a place for certain long-term animal experiments, and there is a place for animal experiments such as the ones that were described yesterday on LAL.

There is also a place for medium term clinical experiments
on well-defined populations - stressed and nonstressed. Some
examples of such experiments have been reported at this meeting. Fomon, who had shown the absolute equivalence of soy
plus methionine with animal proteins in infant formulas,
raised the question of what happens to the serum urea nitrogen
in infants who are fed soy protein isolate to which methionine
was not added. Long-term experiments of use and toleration
like those reported by Scrimshaw in normal adults for a period
of two to four months are important. Or experiments on a special population - the obese, for example. The latter represent a population that sometimes are subjected to higher than
normal protein diets under modified fasting conditions. They
represent an opportunity to determine the tolerance to soy
protein. Or children in school, or premature infants. So
there is the need for ongoing clinical research.

The third need is observations on special population
groups. We will not reach the 10% replacement in a hurry;
it's going to take several years, and it's all going to be in
small increments, and it may be immeasurable over short periods. But there are always some populations that move faster
than others; they must be sought out. I have never been happy
with the state of our information on vegetarians in the United
States. There are other special populations or there are populations that move from one country to the other. Much has
been made about Japanese moving to the United States in other
studies; much has been made of new immigrants to Israel where
major changes in disease patterns have been attributed, in
part, to dietary changes. There ought to be opportunities for
determining the effects, if any, on those who use more soy
products - more rapidly.

The questions about the long-term consequences of putting
soy in the diet -- and this is one of the things that comes
about as you graduate soy into a normal food -- are not different in principle from other questions. For example, what
have been the long-term consequences of doubling the per capita meat intake of the population in the United States in 25
years? This is a far greater change than anything that could
happen from substituting soy protein for 10% of the meat protein. Or the consequences of changing the character and percentage of the simple and complex carbohydrates in the diet;
or something that happened very recently in the grading regulations of meat to permit less fat in quality meat; or increasing the consumption of skimmed milk; or increasing the
consumption of vegetable oil; or a sizeable population in the
United States not being in energy balance; or the percentage
of animal protein rising from 1/2 to over 2/3's of the total
protein. All these represent major changes in this century.

Any one of these can have profound influence on health and
needs to be monitored. Such is a difficult task; it is probably even impossible to assign a clear cut cause-and-effect
relationship; one surely must feel humble in the face of such
complexity. So we have to monitor all these changes as best
we can. We should, likewise, monitor the changes attributed
to soy in the same atmosphere and with the same kind of
approach as we attempt to monitor other major dietary changes.

The next question deals with specifications for soy products. The purpose of specifications is to relate cumulated
nutritional information to a specific product so that the consumer knows what is being bought and regulatory agencies can
provide proper regulations and labeling instructions.

If an unknown material is proposed as a source of protein,
it must be analyzed for nitrogen content and tested in a bioassay to determine whether indeed it merits being considered
as a protein source and whether the further extensive work
needed to explore its biological value can be justified. Once
this is done and equivalence studies are begun, the strategy
should be to develop ultimately a chemical measure that could
suitably describe the desired form. Plant proteins differ
from each other widely in their composition and in their reaction to processing. Hence, a general specification on plant
proteins is well nigh impossible to prepare. Rather, the
strategy should be first to identify the material (e.g., soy
protein isolate, soy protein concentrate, soy protein flour,
cottonseed protein flour, etc.) and then determine the smallest number of measures that would identify the specific product.

Many years ago we prepared specifications of cottonseed
protein flour suitable for human food products such as Incapa-
rina. These included protein content, ϵ-amino lysine content
(a measure of heat-damage) free and bound gossypol contents,
and level of microbiological contamination. No bioassays were
needed to assure protein quality. We knew the product and the
processing, and were able to define it simply and by chemical
measures.

By the same reasoning a specification on soy protein isolate, for example, might include the following:
      a. Protein Content
      b. Measure of heat damage (ϵ-amino lysine content)
      c. Maximum level of trypsin inhibitor
      d. Measure of digestibility
      e. Minimum cystine plus methionine content
      f. Maximum level of microbiological contamination.
I haven't heard anything at the meeting about methods of
measuring the amounts of soy in food products. I think that

sooner or later a good method, and I don't think a good one
exists, must be developed for determining how much soy there
is in a soy-meat combination or in soy-food combinations, because that is the way it's going to be - in combinations. The
regulatory agencies and the consumers would feel much happier
if they knew the proportion in any product. That matter, I
would think, is an item for further research.

The last point that I would like to cover is the social
value of soy in a food economy. This is not an idle matter
because the appreciation of social value changes the climate
within which regulations are developed.

New food products get their start as products of a new
technology. And it may stop at that point: better formulation or appearance or taste or a new form -- an improvement in
hedonic characteristics or hedonic variety -- a social benefit
but a limited one.

When the new product confers an economic benefit, then the
social benefit could be greater, and there could be nutritional overtones. Soy protein - meat combinations are cheaper
than meat alone. For those who desire food on the animal
flesh model, or who think they need it, or in fact need more
protein for special reasons -- for them the soy-containing
foods are an economic benefit that borders on a nutritional
benefit. And in those countries where incomes are rising and
with them expectations for higher quality of life, the ability
to formulate less-expensive products on the animal flesh model
constitutes a real social benefit.

Dr. Hardin spoke about applications to some of the poorer
countries of the world; here too, soy-meat, soy-fish, or soy-
dairy product combinations increase flexibility in dealing
with food problems in the context of poverty and insufficiency
of total food supply.

I would like to mention another nutritional benefit aris-
ing from the introduction of soy-containing foods. This comes
from a reexamination of the concept of variety.

Variety is a fundamental axiom of good nutrition. It was
considered to have maximum value when nutritional information
and knowledge was less available. When we did not know much
about vitamins or where to get them, we proposed variety to
ensure that one could get all the vitamins; and this was really the basis of applied nutrition. Now that we know more
about most of the vitamins and minerals and know where they
are in foods and how to synthesize them away from foods, variety in that sense is less meaningful because there are alter-
natives via the supplements or by fortification. In the case
of unknown micro-nutrients, variety still has meaning; and in
the case of fiber where definite information is lacking on its
nature and role, variety has relevance. How about the relative role of plant and animal protein? We listened to the

interesting paper by Carroll yesterday that raised questions about how much animal protein we might want; or suggested that perhaps there is a range of optimum blends of vegetable and animal protein in a diet.

The ability to interchange animal and plant protein in models that are like meat, sausage, or dairy products raises the concept of variety to a new level because it now allows each individual, depending on that person's idea of his or her own needs, to change the kind of protein in the diet in the way that is considered most desirable.

The forerunner of that blending concept was the variety that inadvertently was introduced to the food economy by the invention of margarine. Then the issue was entirely economic. But now its existence allows one to control better the fatty acid mix depending on needs or assumed needs. This is an element of variety that is important in today's dietary economy.

How about minimizing exposure to toxic materials? This is another great problem. People worry about cancer, they worry about potential carcinogens in foods. They can, if they so wish, reduce their worries by increasing variety in the diet, and this option is made available by new food models which are like existing foods but contain different ingredients. Then one can dilute out any one given ingredient. Therefore, the introduction of proteins of plant origin into familiar models allows the consumer to regulate the proportion of animal and vegetable protein and reduce the level of exposure to any given food source. This confers an important advantage by increasing the consumer's discretion to choose from a wide variety.

The availability of soy products in increasing numbers provides yet another tool for achieving greater flexibility in the diet in the management of nutritional problems. This is the essence of the benefit of nutritional knowledge; and this is an ideal example of a benefit made possible only through advances in nutrition, coupled with progress in food science.

And this potential contribution of soy products to the concept of variety is something about which we should know more.

CONFERENCE SUMMARY

H. L. Wilcke

More than two days of discussion covering a fairly broad
field of the knowledge regarding soybeans in human nutrition
have encompassed a broad, but hardly complete, review. Most
of this consideration has been directed toward the soybean
protein, primarily in the form of the isolates; but also with
some reference to the concentrates and even to the soy flour.
Some of the isolates have been identified, and others have
been referred to simply as the soy protein isolate. The significant point here, though, is that the values and the problems of the use of soy proteins in the human diet have been
discussed, opinions probed; and a much better understanding of
the real potential of the soybean in the human diet has been
attained.

This Conference was planned to meet certain definite objectives, which were presented at the opening of the Conference. It seems appropriate that at the end of the Conference
we should review, briefly, our success or failures in meeting
those Conference objectives.

Obviously, objectives 1 and 4, which were (1) to present
the most up-to-date information possible on the role that soy
protein should fulfill in the human diet and (4) to provide
another forum for the acceleration and improvement of communication of newer and more complex information on soy protein,
have essentially been met by the very fact that the Conference
has been held. Communications have been excellent through
both the presentations and the discussions. Differing points
of view have been expressed and explored, resulting in a better understanding of the meaning of some of the results, and
in pointing out inadequacies in some areas of information that
is available.

The types of soybean protein products available have been
quite clearly defined, based upon their protein content and
the general methods of preparation; and these definitions have
been quite· generally accepted and understood.

The recent meetings of the committees on the Codex Alimentarius have provided better understanding and agreement among nations on the philosophies of regulatory practices. Primary differences appear to be in the areas of whether soy products should be supplemented to bring PER's to equivalence with meat products or other products they may replace, or whether the soy products should be characterized and described as they are offered, providing full information to the consumer and to the food industry, permitting proper use to be determined by the composition of each product.

The existence of different classes of soy products available to the food industry, varying from flour to concentrate to isolate, all in various physical forms but produced by differing processing procedures, and with varying chemical composition, makes it imperative that the type of soy protein used in experiments be identified. This, of course, is true with all research work and not only with work in the area of soybean products.

It has been suggested that the concept of functionality be broadened to include nutritional and physiological function of the proteins, and information presented regarding the unique properties of soy protein in affecting cholesterol metabolism has supported this concept. This also emphasizes the need to evaluate functionality in the final food product rather than in the ingredient per se. The concept of functionality has become increasingly vague and general, and it has been suggested that it would be more appropriate to use a modifier such as color, flavor, texture, or nutrition with the term "functionality." This certainly would identify the characteristic for the intended use.

The second objective of this Conference was to stimulate interest among both the public and private sectors in pursuing research programs directed toward the determination of the proper place of plant proteins in the human diet. The results presented in actual feeding programs for infants, children, and young adults clearly show that soy protein is a much better protein nutritionally for humans than would have been predicted on the basis of rat assays; and that it is extremely difficult, if not impossible, to demonstrate a methionine deficiency, particularly when the soy is included in the diet at levels approaching adequacy, such as 6% of the calories, or somewhat higher. No evidence of allergenic reactions has been demonstrated in these experiments. Particular emphasis has been placed on this point through follow-up examinations and studies several years after infants had been fed on diets high in soy protein. No untoward effects were reported. This is in direct contrast to some unsubstantiated reports from other countries that some allergenic response might be involved with certain types of soy products.

The work reported on the nutritional effects of soy have been for the most part rather short-term in nature. Longer term studies seem to be indicated to verify the results obtained to date and to bolster the confidence of the consumer. The results reported here should encourage additional research, and certainly there should be more than one source of data for the various age groups; and the work should be extended to older age groups as well. For, as pointed out in this Conference, protein quality is important to the nutrition of all ages.

Biologically active factors, some favorable and some which may be detrimental, have been reported in the literature. As was pointed out, this is not a phenomenon peculiar to the soybean, but is characteristic of almost any food substance. Of the factors discussed for the soybean, the role of phytate appears to be the least clarified. The evidence indicates that this compound may have an effect on the availability of some of the trace minerals - notably zinc, and perhaps iron, and that calcium does seem to modify the effects. However, it is accepted that there are factors other than phytate affecting the availability of these trace minerals, among which fiber or cellulose have been suspected, although the role has not been clearly delineated. Research is needed to explain differences in availability of these trace minerals because, by definition, they are present in low concentrations, and decreased availability may create marginal, if not frank deficiencies.

Such factors as trypsin inhibitors, hemagglutinins, goitrogens, and lysinoalanine have been demonstrated to be of no nutritional significance, or may be reduced to inconsequential levels by proper processing. In spite of this, interesting unanswered questions remain in regard to pathways of biological activity when these factors are present, or the exact levels which might produce threshold effects; and better methods of inactivation may be desirable in some cases, because of the possibility of detrimental effects on nutritional quality due to the inactivation process itself if applied in excess.

The research demonstrating lower plasma cholesterol levels when the protein of the diet is soybean, or other proteins of plant origin, as contrasted to levels existing when the source of protein is of animal origin, is extremely interesting and must be pursued further. The consequence of this effect, i.e., less atherosclerotic lesions, is extremely important. The fascinating field of the effects on cholesterol metabolism should attract the attention of research to explain more fully the mechanism of this effect, the specific components involved, and perhaps how these effects may be enhanced.

The third objective of this Conference was to stimulate

interest in better methods of evaluating the quality of proteins for human nutrition. Clearly, the need is to define the problems which must be attacked. Too long, the efforts have been to improve methods of evaluating protein without a clear purpose for evaluation other than to determine the effects on growth, or maintenance, or both.

The objectives might be divided into two areas:

One - simple, but reproducible methods which will determine the effects of treatment of the protein products, whether through processing, environmental effects, or others. The method, or methods, would demonstrate differences in the nutritional ranking of various proteins, and would not be concerned with absolute measurements of nutritional value.

The second would be methods of assessing digestibility or biological availability of the amino acids of the protein. This would necessitate standardization of methods of analysis, and perhaps new methods which would provide more precise quantitative results with less probability of destruction of amino acids. This would also, a priori, require more precise determination of the requirements of the human for the essential amino acids at various stages of growth and development, including senescence.

Methods for evaluating the nutritional quality of protein, based on rat assays, do provide much useful information. PER, which is an accepted method for regulatory purposes, does provide information on ranking of proteins. There is still need for standardization of methodology to produce more uniform results from laboratory to laboratory. However, this method, as well as other methods based on rat assays, does leave much to be desired in actual evaluations of proteins for human nutrition. The need for evaluation at more than one level of protein is becoming increasingly apparent, and it has been proposed that levels just above and just below the accepted protein requirement level be used. Thus, the concept of evaluation of a proper use level, rather than at suboptimal levels, seems to be gaining much more support. There remain questions to be resolved between the various modifications of slope ratio, NPU, and PER, and other methods.

Another concept that has been introduced is the need for differentiation of plant proteins in our thinking about evaluation. We do not lump all animal proteins into one group - we consider them as fish, meat, milk, eggs, etc. It is particularly important that regulatory agencies make this distinction in order to develop definitions and regulations that are clear, appropriate to the specific type of product, and meaningful to both the food industry and the consumer. As more plant protein products become available, isn't it going to be necessary to look at them separately, as individual and different entities?

The need for greater sophistication in scoring proteins, based upon amino acid makeup of the protein, is becoming more apparent, and the panel has recommended further study and development.

Thus, it is apparent that our speakers and our discussants have addressed the objectives of this Conference quite effectively. Certainly not all of the questions have been answered completely. This would not be possible. Perhaps not all of the right questions have been asked. It is possible that not all the objectives that should be covered in a conference of this type have been stated. However, a very effective presentation of the data, the interpretations, the penetrating questions that have been raised, and the intense interest that has been displayed throughout the program all are evidence that this program has brought together a great deal of objective information regarding the value of this good food product. The real impact of this Conference will be determined by the number of new research programs, or intensified programs that will be initiated in the next year or two. This will provide evidence of whether interest is real.

Explanation of trademarks used in this publication:

Edi-Pro A, Edi-Pro N, Supro 710, and *Supro 620* are isolated soy proteins trademarked and produced by the Ralston Purina Company, St. Louis, Missouri.

TVP is a textured soy flour produced by the Archer Daniels Midland Company of Decatur, Illinois.

Sobee is an infant formula produced with soy flour as the protein source and trademarked and produced by Mead Johnson, Evansville, Indiana.

Pro Sobee is an infant formula produced with isolated soy protein as the protein source and trademarked and produced by Mead Johnson, Evansville, Indiana.

Promosoy 100 is a soy protein concentrate trademarked and produced by the Central Soya Company of Ft. Wayne, Indiana.

Supro 50-4 is a textured soy flour trademarked and produced by the Ralston Purina Company of St. Louis, Missouri.

KEYSTONE CONFERENCE
REGISTRANTS

Dr. Richard P. Abernathy
Purdue University
2600 Newman Road
West Lafayette, IN   47907

Dr. James Adkins
Howard University
2820 Henderson Court
Wheaton, MD   20902

Dr. Aaron M. Altschul
Georgetown University
2233 Wisconsin Avenue, N.W.
Washington, DC   20007

Mr. Jack J. Anton
Ralston Purina Company
Checkerboard Square
St. Louis, MO   63188

Dr. Philip Aines
Pillsbury Company
311 Second Street, S.E.
Minneapolis, MN   55414

Dr. Robert R. Baldwin
Morton Frozen Foods
P. O. Box 7547
Charlottesville, VA   22906

Dr. John Benson
Ross Labs
625 Cleveland Avenue
Columbus, OH   43216

Dr. Jeremy A. Blake
Syntex Research
3401 Hillview Avenue
Palo Alto, CA   94304

Dr. C. E. Bodwell
ARS-USDA Room 313
Building 308
BARC-East
Beltsville, MD   20705

Dr. Ricardo Bressani
INCAP
Apartado Postal 1188
Guatemala City, Guatemala

Mr. William B. Brew
Ralston Purina Company
Checkerboard Square
St. Louis, MO   63188

Dr. Mahlon A. Burnette, III
Grocery Manufacturers of
  America
1010 Wisconsin Avenue, NW
Suite 800
Washington, DC   20005

Dr. Ken K. Carroll
University of Western Ontario
London, Ontario, Canada
N6A561

Mr. John C. Colmey
ITT Continental Baking
  Company
P. O. Box 731
Rye, NY   10580

Dr. David A. Cook
Mead Johnson Research Center
2204 West Pennsylvania Avenue
Evansville, IN   47721

Dr. Ralph A. Damico
The Procter & Gamble Company
P. O. Box 39175
Cincinnati, OH   45247

Mr. Alain Decock
Ralston Purina Company
52 Rue Des de Ponts de
  Comines
Lille, France

Dr. Charles W. Deyoe
Kansas State University
Shellenberger Hall
Manhattan, KS 66506

Dr. Robert C. Doster
Carnation Company
8015 Van Nuys Boulevard
Van Nuys, CA 91412

Dr. Neil Doty
General Nutrition Corporation
Box 349
Fargo, ND 58102

Dr. John W. Erdman, Jr.
University of Illinois
567 Bevier Hall
Urbana, IL 61801

Dr. Samuel J. Fomon
University Hospital
University of Iowa
Iowa City, IA 52242

Dr. Dick H. Forsythe
Campbell Soup Company
Campbell Place
Camden, NJ 08101

Mr. Irwin Fried
U.S. Department of
  Agriculture
Room 202, Annex Building
Washington, DC 20250

Mr. Henry P. Furgal
Miles Labs
601 East Algonquin Road
Schaumburg, IL 60195

Dr. L. Ross Hackler
New York State Experiment
  Station
Cornell University
Geneva, NY 14456

Dr. Clifford M. Hardin
Ralston Purina Company
Checkerboard Square
St. Louis, MO 63188

Dr. Alfred E. Harper
Department of Nutritional
  Sciences
University of Wisconsin
Madison, WI 53706

Dr. Robert Harper
Milnot Company
P. O. Box 190
Litchfield, IL 62056

Mr. Ronald D. Harris
Anderson Clayton
3333 North Central Expressway
Richardson, TX 75080

Mr. Vaughn Hatch
Coco Cola
497 Plum Street
Atlanta, GA 30313

Mr. Paul H. Hatfield
Ralston Purina Company
Checkerboard Square
St. Louis, MO 63188

Dr. Lavell M. Henderson
Biochemistry Department
University of Minnesota
St. Paul, MN 55108

Dr. Daniel T. Hopkins
Ralston Purina Company
Checkerboard Square
St. Louis, MO 63188

Dr. Marc Horisberger
Nestle Products
Case Postale 1009
Lusanne, Switzerland

Dr. J. Edward Hunter
Procter & Gamble Company
6071 Center Hill Road
Cincinnati, OH   45224

Dr. James M. Iacono
U.S. Department of
   Agriculture
14th and Independence, S.W.
Washington, DC   20250

Dr. Goro Inoue
Tokushima University
3 Kuramoto-Cho
Tokushima 770 Japan

Dr. G. Richard Jansen
Department of Food Science
Colorado State University
Fort Collins, CO   80523

Mr. Al Julius
Iowa State University
Kildee Hall
Ames, IA   50011

Dr. Constance Kies
College of Home Economics
University of Nebraska
Lincoln, NE   68503

Dr. Charles W. Kolar
Ralston Purina Company
Checkerboard Square
St. Louis, MO   63188

Dr. Irvin E. Liener
Biochemistry Department
University of Minnesota
St. Paul, MN   55108

Mr. Ted R. Lindstrom
General Foods Technical
   Center
250 North Street
White Plains, NY   10625

Dr. Grace S. Lo
Ralston Purina Company
Checkerboard Square
St. Louis, MO   63188

Mr. Haines Lockhart
The Quaker Oats Company
617 West Main Street
Barrington, IL   60010

Dr. J. B. Longenecker
University of Texas
Austin GFA 115
Austin, TX   78712

Mr. Edmund H. Lusas
Food Protein R&D Center
Texas A&M University
College Station, TX   77843

Dr. Ronald L. Madl
Ralston Purina Company
Checkerboard Square
St. Louis, MO   63188

Dr. Boaz A. Mafarachisi
Kellogg Company
235 Porter Street
Battle Creek, MI   49016

Dr. Elly T. Margolis
Seyforth Labs, Inc.
3118 Depot Road
Hayward, CA   94545

Mrs. Wilda Martinez
U.S. Department of
   Agriculture
Room 224, Building 005
Beltsville, MD   20705

Dr. John McLaughlan
Department of National Health
   and Welfare
Tunney's Pasture
Ottawa 3, Ontario, Canada

Dr. Max Milner
Massachusetts Institute of
   Technology - 20A-224
Cambridge, MA   02139

Dr. Julian Mincu
Nutrition Institute
Bucharest,
Rumania

Dr. Robert E. Moreng
College of Agricultural
   Science
Colorado State University
Fort Collins, CO   80523

Dr. Charles V. Morr
3125 Cortina Drive
Colorado Springs, CO   80918

Dr. John F. Mueller
St. Luke's Hospital
601 East 19th Avenue
Denver, CO   80203

Dr. Robert O. Nesheim
The Quaker Oats Company
617 West Main Street
Barrington, IL   60010

Dr. Boyd L. O'Dell
University of Missouri
322 Chemistry Building
Columbia, MO   65201

Dr. Ragnar Ohlson
Karlshamns
AB Karlshamns Oljefabriker
S-292 00 Karlshamn, Sweden

Dr. Winston S. Ogilvy
Mead Johnson & Company
2404 Pennsylvania Avenue
Evansville, IN   47721

Dr. Peter L. Pellett
University of Massachusetts
Food Science and Nutrition
Amherst, MA   01003

Dr. Richard A. Phelps
Anderson Clayton
P. O. Box 2538
Houston, TX   77001

Dr. B. P. Poovaiah
Shaklee Corporation
1922 Alpine Way
Hayward, CA   94540

Dr. Peter E. Priepke
Campbell Institute
Campbell Place
Camden, NJ   08101

Dr. Joe J. Rackis
U.S. Department of
   Agriculture
1815 North University Avenue
Peoria, IL   61604

Dr. Stan H. Richert
Ralston Purina Company
Checkerboard Square
St. Louis, MO   63188

Mr. Edwin J. Rigaud
Procter & Gamble Company
6071 Center Hill Road
Cincinnati, OH   45224

Dr. S. J. Ritchey
Virginia Polymer Institute
   State University
Blacksburg, VA   24061

Dr. Howard Roberts
Food and Drug Administration
200 C. Street, S.W.
Washington, DC   20204

Mr. Leonard H. Roberts
Ralston Purina Company
Checkerboard Square
St. Louis, MO   63188

Dr. Daniel Rosenfield
Miles Labs, Inc.
1127 Myrtle Street
Elkhart, IN   46514

Dr. John T. Rotruck
The Procter & Gamble Company
P. O. Box 39175
Cincinnati, OH   45247

Dr. Max Rubin
University of Maryland
Jull Hall, Room 2110
College Park, MD   20742

Dr. Ken W. Samonds
University of Massachusetts
06 Dana Street
Amherst, MA   01002

Dr. Sidney Saperstein
Syntex Research
3401 Hillview Avenue
Palo Alto, CA   94304

Dr. Lowell D. Satterlee
Food Science and Technology
University of Nebraska
Lincoln, NE   68583

Mr. H. Leroy Schilt
Ralston Purina Company
Checkerboard Square
St. Louis, MO   63186

Dr. Gustav Schonfeld
Washington University
4566 Scott Avenue, Box 8046
St. Louis, MO   63110

Dr. Nevin Scrimshaw
Massachusetts Institute of
  Technology
77 Massachusetts Avenue
Cambridge, MA   02139

Dr. Fred R. Senti
American Society for Exp.
  Biology
9650 Rockville Pike
Bethesda, MD   20014

Dr. Rose Ann Shorey
Department of Home Economics
University of Texas - Austin
  GEA 115
Austin, TX   78712

Mr. Ronald Simpson
Standard Brands
Betts Avenue
Stamford, CT   06904

Dr. Keith J. Smith
American Soybean Association
Box 158
Hudson, IA   50643

Dr. Fred H. Steinke
Ralston Purina Company
Checkerboard Square
St. Louis, MO   63188

Dr. Barbara J. Struthers
Ralston Purina Company
Checkerboard Square
St. Louis, MO   63188

Dr. Marian F. Swendseid
School of Public Health
University of California
Los Angeles, CA   90024

Dr. William Tallent
U.S. Department of
  Agriculture
1815 North University Street
Peoria, IL   61604

Dr. Hitoshi Taniguchi
Fuji Oil Company, Ltd.
6-1 Hachimancho
Osaka, Japan (T 542)

Mr. A. Terral
French Ministry of
  Agriculture
Legislation Meat Products
Paris, France

Dr. William Thomson
Ross Labs
625 Cleveland Avenue
Columbus, OH   43216

Dr. R. M. Tomarelli
Nutrition Department
Wyeth Labs
Radnor, PA   19087

Dr. Benjamin Torun
INCAP
Apartado Postal 1188
Guatemala City, Guatemala

Dr. Piet Van Stratum
Unilever Research
P. O. Box 114, 3130 AC
Vlaardingen, The Netherlands

Dr. John E. Vanderveen
Food and Drug Administration
200 C. Street, S.W.
Washington, DC   20204

Dr. Doyle H. Waggle
Ralston Purina Company
Checkerboard Square
St. Louis, MO   63188

Dr. David F. Walsh
General Nutrition Corporation
1301 Thirty-Ninth Street
Fargo, ND   58102

Dr. Kenneth D. Wiggers
Iowa State University
313 Kildee Hall
Ames, IA   50011

Dr. Harold L. Wilcke
Ralston Purina Company
Checkerboard Square
St. Louis, MO   63188

Dr. Dean Wilding
Kraft, Inc.
801 Waukegan Road
Glenview, IL   60025

Dr. James L. Williamson
Ralston Purina Company
Checkerboard Square
St. Louis, MO   63188

Dr. Walter J. Wolf
U.S. Department of
  Agriculture
1815 North University Street
Peoria, IL   61604

Dr. Allison V. Zezulka
University of Texas
1100 Halcombe Boulevard
Houston, TX   77025

Dr. Ekhard F. Ziegler
University of Iowa
Iowa City, IA   52242

INDEX